© Steven Cotton

All rights reserved. No part of this book may be used or reproduced in any manner without prior written permission except in the case of brief quotations in critical articles or reviews.

Inevitable:

The Future has Already Happened

Inevitable

The Future has Already Happened

STEVEN COTTON

To all whose hunger and thirst for knowledge is
Unquenchable

Table of Contents

INTRODUCTION 6

CHAPTER ONE: PARADIGMS LOST 8

CHAPTER TWO: THREADING THE NEEDLE 17

CHAPTER THREE: THE JOURNEY BEGINS 33

CHAPTER FOUR: TANGLED WEB 41

CHAPTER FIVE: THE DISCOVERY 55

CHAPTER SIX: STRANGE BEDFELLOWS 71

CHAPTER SEVEN: PRELUDE 75

CHAPTER EIGHT: PORTAL TO THE UNKNOWN 83

CHAPTER NINE: MORE SECRETS 94

CHAPTER TEN: INTERLUDE 103

CHAPTER ELEVEN: CONVERGENCE 118

CHAPTER TWELVE: NOOKS AND CRANNIES 130

CHAPTER THIRTEEN: THE MOTHER LODE 142

CHAPTER FOURTEEN: KEEPING TRACK 160

CHAPTER FIFTEEN: THE ANT SWALLOWS THE ELEPHANT 172

CHAPTER SIXTEEN: THE CAT THAT ROARED! 177

CHAPTER SEVENTEEN: PARADOX 198

CHAPTER EIGHTEEN: INEVITABLE 208

CHAPTER NINETEEN: PREDESTINATION REDUX 220

CHAPTER TWENTY: STRANGER THAN TRUTH 231

CHAPTER TWENTY-ONE: ALTERED STATES 252

CHAPTER TWENTY-TWO: DESTINY 262

Introduction

Inexplicable anomalies crop up in our lives when we become receptive to their power. A time comes when we are challenged to confront our complacency and submit to the possibility that our comfortable worldview is no longer tenable. We are then forced to see the world in a new light, to discard our comfortable beliefs, and envision a world evolving from the familiar to the extraordinary.

The science presented here introduces the reader to a new way of viewing the cosmos and our place in the evolutionary history of the universe. Time paradoxes, quantum entanglement, and parallel universes provide a backdrop for exploring a range of scientific speculations about the true nature of time and the ultimate meaning of reality.

Scientific discoveries in physics, genetics, and quantum reality are introduced, encompassing advanced techniques that can only have been achieved in a future world beyond the reach of the present. As these discoveries unfold, the reader begins to realize conventional wisdom is suspect — that what we believe, what we have come to accept without question, is subject to revision.

Our investigation follows a scientific approach and the reader is exposed to the thoughts, actions, and psychological machinations of a trio of scientists as they attempt to uncover the origin of an ancient Egyptian tablet containing an equation from quantum mechanics. Several hypothetical scenarios are considered to explain the tablet's appearance and the reader is challenged to comprehend the many possibilities introduced.

Chapter One: Paradigms Lost

During his tenure at Cairo University in Giza, Egypt Dr. Syed Azad has introduced many innovative mathematical concepts into the field of Egyptology. He calls me on occasion seeking my opinion of his latest research. We have been keeping in touch on a collaborative basis since our first meeting a few years ago at a conference in Paris, France. The symposium was designed to explore the link between mathematics and ancient architectural structures. As a professor of mathematics and physics at Stanford University, my role is to evaluate the soundness of the math as it applies to various theories under consideration.

When Syed calls me early Sunday morning, I prepare myself to listen to another one of his wild ideas. He does have his brilliant moments but they tend to come to him when I am in deep slumber and he is wide awake. This time is different.

"Syed, it's good to hear from you," I say, hoping the distraction will brighten up the beginning of a routine sabbatical from the university.

"Jake, my good friend. How are you? Still getting paid to confuse your students?" Syed speaks with a light accent in well-practiced English. His sense of humor is brisk but good-natured. I only take him seriously when he does.

"Listen, Jake, I have come across a quite remarkable find." Syed pauses for a moment, gathering his thoughts. "I have a tablet in my possession which is not like anything I have ever seen. The inscription contains an equation. It is the Schrödinger equation you use in your work with quantum physics."

I am familiar with the equation. It is a partial differential equation describing how a quantum state changes over time. The Austrian physicist Erwin Schrödinger derived the equation almost a century ago. Why would it be etched into an ancient Egyptian tablet? How did it end up in Dr. Azad's possession?

"Where did this tablet come from?" I ask, thinking there must be some mistake. It must be a forgery, a hoax perpetrated upon an unsuspecting public. Crop circles come to mind. "Syed, have you verified the tablet's origin? It couldn't possibly be authentic—"

Before I can continue expressing my incredulity, Syed interrupts. "We have verified the date to the beginning of the Early Dynastic Period, about three thousand BC, but the exact location remains uncertain. I believe the location of the find is near the ruins in the town of Mit Rahina just south of Cairo."

My mind refuses to wrap itself around such an absurdity; and yet, when I consider the source, Dr. Syed Azad—a world-renowned Egyptologist unsurpassed in reputation and authority (with the possible exception of Dr. Zahi Hawass)—I can't simply dismiss the possibility out of hand. Syed has all the bona fides: A prodigious work product on numerous digs; extensive research and innovation in both the discovery and deciphering of Egyptian hieroglyphics; numerous publishing credits; and, perhaps most important, a well-deserved reputation in the Egyptology community.

Dr. Azad is well into his sixties but has retained a youthful appearance he attributes to "clean living" (although he enjoys a sip of his favorite lager on occasion). He often slouches but when he stands up straight, he is all of six-foot. (When we first met in person, I thought he was about my height, five-foot-nine inches.) He is relatively trim and fit and his salt-and-pepper hair lends an air of sophistication. His interpersonal skills lack refinement but he still manages to make friends easily and his well-honed intelligence, Mediterranean good looks, and a quick wit make it easier to forgive his slights when he often speaks out of turn.

I tell Syed there is too much to process. The implications, if born out, would usher in a paradigm shift unprecedented since Einstein's theories of Relativity in the early twentieth century. I suggest to Dr. Syed we keep things under wraps for now telling him I will be getting back in touch as soon as possible. I can't think of anything else to do at the moment. Figuring out where to go from here will take careful planning, skill, and time.

Geography prevents Syed and me from becoming close friends but we enjoy an occasional collaboration and I always look forward to hearing from him. He first contacted me about a year before the Paris conference. He had asked me about a paper I had submitted to the International *Journal of Science and Mathematics*. In the article I attempted to debunk the idea that advanced mathematics was involved in the construction of the pyramids in Egypt, using Khufu's Great Pyramid as a focal point of my analysis. My paper dealt with the geometry embedded in the design of the Great Pyramid. Although my approach was not new (I did not introduce any groundbreaking insights into the never-ending debate about the construction of the pyramids), my use of mathematical proofs intrigued Dr. Azad and he contacted me to discuss the implications of my analysis.

My area of expertise lies well outside the study of Egyptology. Pyramidal design and structure, however, are fair game for any enterprising mathematician and I thought I could provide at least a modicum of a fresh perspective on the subject. Since collaborating with Dr. Azad my appreciation for and understanding of Egyptian architecture has increased exponentially. So much so I now understand how unequipped I was to have attempted a contribution to this field when I did.

Syed could have shared his discovery with any number of associates in the local, scientific community, but he chose me. I think I know why but don't want to jinx it by asking. He would have considered the implications of his discovery and knows that any breach of confidentiality would be catastrophic. He had told me previously that he could not count on discretion among his colleagues. Rivalry often leads to in-fighting among his associates as they attempt to discredit each other's discoveries. Syed wanted no part of it.

Confidentiality is important, very important. But the real reason Syed contacted me is more fundamental: He likes me and me him. Since our first meeting in Europe three years ago, we hit it off spectacularly (Drs. Azad and Banner quickly becoming Jake and Syed). Our thoughts resonate on almost every subject. In addition to science and mathematics, we also share a good deal of interest in literature and classic movies. Syed has acquired much of his understanding of American culture from watching older American films. His favorite is *Casablanca* starring Humphrey Bogart and Ingrid Bergman. He once told me after a long discussion regarding who knows what, "Jake, I think this is the beginning of a beautiful friendship."

From childhood, I have manifested symptoms of what I generously call *shadow syndromes*. I have always had a bit of OCD, a smidge of hypochondria, and, just to top it off, a dollop of paranoia. All essential "skills" I use effectively in my avocation.

Assessing the current impact of Syed's discovery, I realize immediately that, if corroborated, the entire history of science as we know it will be thrown into question. If a major tenet of quantum mechanics had been laid down on a tablet in ancient Egypt thousands of years ago, that could only mean one thing: Someone or something had scooped Schrödinger by five millennia! How could this discovery have remained dormant for so long? How could this tablet document, beforehand, a major scientific development of the twentieth century? Are there other tablets of a similar nature? These questions percolate in my mind as I try to wrestle with the consequences of the discovery and formulate a game plan for future action.

After giving the matter careful thought, I realize there is only one thing to do: Hit the "gym"! I converted a bedroom into a makeshift fitness center a few years ago when I realized my physique needed a little fine-tuning. Before installing my workout room, I would jump rope in the morning and do a few reps with live weights. Funny how old habits never die—I still jump rope and work with live weights; the new gym now augmenting my morning routine. In addition to my height, my genetic allotment has given me green eyes and a receding hairline, "prematurely grey." I opted for glasses instead of contacts or eye surgery. All in all, I think I still look relatively young at fifty-eight years of age but don't want to give the reaper any advantage. Besides, working out is therapeutic and allows me time to meditate while keeping fit.

After completing my exercise routine, I take a quick shower, ready to tackle the mystery at hand. I decide to approach the matter as I would any scientific problem. I will first gather the facts in greater detail and formulate various hypotheses to account for the tablet's sudden appearance in Egypt at this time in history (Why was it only now discovered?). I will then need to plan for and anticipate any future repercussions that will inevitably follow public awareness of the find. This approach sounds logical and has stood me well in my academic career. And yet . . . somehow, I can't shake the unsettling feeling that what I don't know will become more important as I wrestle with the inexplicable nature of this impossible discovery.

I very seldom use checklists; they tend to weaken the mind. But this is an exception. I can't afford any mistakes. Before following up with Syed I need to be certain I haven't missed anything. After jotting down some handwritten notes, I summarize them in the form of questions to help me organize my thinking:

1. How, precisely, did Dr. Azad acquire the tablet? To use legal terminology: What was the *chain of custody*?
2. Is the tablet a *standard* plate? Does it have the normal dimensions associated with a find of this period? Is its material composition similar to others found in the same location?
3. Are there any other markings on the tablet?
4. Is anyone else aware of the find?

5. Are the markings on the tablet consistent with the type of tools used during the period? (Do the grooves in the plate and other forensic evidence correlate with the historical timeframe in question?)
6. Is there any way to locate the person who had originally come into possession of the plate?
7. What specific method was used to authenticate the age of the artifact?
8. Has Syed or any of his colleagues ever encountered anything . . . anything, out of the ordinary in their previous excursions in the field that might shed some light on this recent find?
9. Is there any possibility of visiting the original site where the plate was presumed to have been discovered?
10. Will Syed be in a position to ask for an extended leave from University if it becomes necessary to pursue the investigation further?

After completing the checklist, I decided to take a walk to clear my head. Living in a faculty housing area within a mile of the campus has its advantages. I soon arrive at one of my favorite hangouts—a *Starbucks* café just beyond the faculty club. I order my usual, a *Grande Americano* with two sugars and a little cream. I'm lucky to find a spot by the window. Just as I am about to relax and enjoy my brew, I see Linda Cooper sitting alone in a far-off corner. I am surprised to see her here on a weekend. Linda and I have a close, professional relationship that began soon after she joined the faculty about five years ago.

Linda is a tall, beautiful brunette with shoulder-length hair and eyes that put paid to the expression, "dark as the deep blue sea." When she wears heels, she can look me squarely in the eyebrows! She teaches genetics and molecular biology and is currently working on a hypothesis dealing with directed genetic mutations to overcome both mental and physical disabilities. I like to think I have a keen intellect but when she gets into the weeds explaining alterations in protein synthesis and the reassignment of "junk DNA," my concentration tends to wane. Linda, on the other hand, can grasp any topic; always finding a way to listen and contribute. She is a good sounding board and now seems like the ideal time to pick her brain and see what she might think about *my* discovery.

  I can't come right out and tell Linda the whole story, however. That would violate confidence and push the story along before it is properly vetted. As I walk over to greet her it hits me: I'll just spin a scenario; tell her I had been daydreaming during my walk and ask her what she thinks about my latest goofy idea. She is used to my meanderings and we both enjoy the friendly banter.

  "Hey," I say, this being our customary greeting.

  "Hey" she replies. "I thought you made plans. What are you doing here?"

  "Oh, the plans are made. I just haven't executed them yet." As I speak, I realize I haven't sat down yet. I am excited about the prospect of ferreting out the details of what Dr. Azad has uncovered. I sit down and, after exchanging a few pleasantries, begin to spin my tale.

  "What if our lives and everything we have experienced have been programmed? I know it isn't a completely new idea; physicists have broached the subject before. But what if there is conclusive evidence, independently verified and corroborated?"

  After letting the gist of what I proposed to Linda sink in, she gets to work. "If everything we experience has been programmed, then the very act of questioning this possibility has also been programmed. And if this is so, we must then conclude that inquisitiveness itself is an illusion. We would in effect be nothing more than sophisticated machines programmed to believe we have volition and mental faculties capable of self-discernment. The very act of suspicion would itself be suspect."

  "That's amazing!" I say. "How do you do that?"

"Modus ponens" she replies, looking at me in that playfully self-assured way she does when she knows she has me flummoxed.

"What does that mean? You're not going Latin on me, are you?"

"P implies Q. If P is true, Q must be true. It's basic logic."

"Okay," I say, "but what if our history isn't predetermined so much as simply documented in advance? Or certain key events are known beforehand?" I toss out this modified version of my thought process to see if she catches the nuance.

"That would imply foreknowledge as well," she says. "Either way, some form of intelligence must be operating; an entity or system capable of formulating or even orchestrating history. We would essentially be out of the loop and at the mercy of forces—not only beyond our control—but *controlling* us. Not a pretty picture as far as I am concerned."

We chat for a while longer, discussing current events and sundry items, before I head back home. A leisurely walk along Lagunita drive would normally have calmed any lingering anxieties I have about this whole affair, but not today. My *shadow syndromes* are beginning to make inroads on what is left of my sense of certainty and confidence about the world and my place in it.

Linda's assessment is irrefutable. But, if true—and using her "modus ponens" logic—the discovery by Dr. Azad and my subsequent involvement would also be inevitable. The whole thing is too crazy to take seriously . . . and yet, there *is* evidence. If this discovery becomes public knowledge it will turn the whole scientific community into a state of bedlam and I don't want to be the headmaster of the asylum. Maybe I can convince Dr. Azad to destroy the tablet and pretend it never happened. But wait! If I do that, then that outcome too would be . . .

The way I had presented the scenario to Linda left little room for an alternative conclusion. If *everything* is predetermined then it logically follows that all of our actions and even history itself is also "written in stone." Dr. Azad's discovery however is singular—an individual finding depicting foreknowledge of a mathematical construct. This single artifact does not encompass all of human history, far from it. Who knows, perhaps the tablet is one among many. Further discoveries might shed light on this otherwise inexplicable mystery. I resolve to contact Dr. Azad tomorrow to explore this possibility.

Chapter Two: Threading the Needle

Monday morning seems to have arrived earlier than usual; my normal sleeping pattern disrupted by the anticipation of what may lie ahead. Usually a sound—very sound—sleeper, I barely manage to capture sufficient rem time. Perhaps my morning routine will recalibrate my senses: Coffee on the patio, latest updates on the news and financial markets, and a small bowl of oatmeal, with cinnamon.

The fresh air is invigorating—crisp, cool, and just the right touch of chill. It is barely coming up at six a.m. West Coast time. What time would that make it in Cairo? Sometimes when I need to look something up on the net, I remember the line from one of those nondescript movies to "Google it!" So, I Google it. Six a.m. here would be three p.m. Cairo time. I should be able to reach Syed at home shortly.

If Syed and I decide to track down the tablet's origin, this would be the ideal time of year to do so. The weather in Egypt during early March is just entering the "Goldilocks" zone—neither too warm nor cold but just right. I will need to cancel my vacation plans; the Caribbean can wait. Oh well, there goes my deposit. What if the trip is a bust and we make no further discoveries? What if everything turns out to be a hoax after all?

My attempt at normalizing my day with familiar routines isn't working. My mind skips from one thought to the next. Sometimes when I need to regain my concentration, I play chess "against the computer." Chess programs range from easy to difficult with some "difficult" levels being *more difficult* than others. I pull up my favorite program (playing white), choose the "more difficult" level, and began my strategy: Pawn to king 4, bring out the knight, and so on. You can't go wrong with the *Ruy Lopez* opening (well, you can, but it is a staple). I begin to develop my pieces and quickly get into serious trouble. The problem with computer chess programs is that they don't think (and therefore make mistakes); they simply execute an avalanche of algorithms. No emotion. No second thoughts. No feelings.

Are we nothing more than an amalgamation of sophisticated algorithms, executing one module after another, all the while believing that we are in control? The author Tor Norretranders, in his book *The User Illusion: Cutting Consciousness Down to Size* seems to be saying as much. Current updates in the field of neurology and brain science tend to corroborate his theory.

Anyone acquainted with the science of neurology is aware of Benjamin Libet's work in the field of human consciousness and the neuroscience of free will. During the seventies, he researched physiology at the University of California, San Francisco, where he attempted to determine the connection between the activation of specific sites in the brain and somatic sensations. Libet's last paper published in *Brain* in 1991 confirmed—in the words of Tor Norretrander—that "consciousness is a fraud." Libet's work suggests that electrical impulses in the brain precede conscious awareness. Essentially, our intentions to perform a specific action are initiated subconsciously, followed by our conscious awareness of such intentions. Anyone confused and believing this to be "ass-backward" is to be congratulated—it is a clear sign they are paying attention!

The rabbit hole is even deeper than we think, with many twists and turns. *The Matrix*, a nineteen ninety-nine American-Australian film starring Keanu Reeves, depicts a future dystopian world (Why do futuristic sci-fi movies always show the underbelly of civilization?) where machines create an artificial reality. Neo (Keanu) plays the protagonist who discovers the true nature of the simulated reality program and ends up leading a struggle against the forces of darkness. Cypress is a crewmate of Neo who, although aware of the make-believe nature of the matrix program, nevertheless opts to betray Morpheus (captain of a hovercraft living in the last remaining human dwelling) selling out for a life of fiction inside the matrix. If we are to believe the unfolding revelations of neuroscience we just as surely live in a matrix, even if it is one of our makings and not imposed from the outside.

What is believable is not necessarily true. We each have our very own, personal, model of reality. As odd as it may seem each of our models, even if inconsistent with another's, works for each of us . . . until it doesn't. When this happens, we are forced to either modify or abandon our current model and continue with an adjusted or completely new one. The Ptolemaic model of our solar system—epicycles and all—works just fine. It is a far cry from the updated Copernican model but I am convinced we could still use it, with perhaps some minor tweaking, to land an astronaut on the moon. It may not fully explain the precession of Mercury (that was left to Einstein to explain) but it'll do for our purposes.

Dr. Azad's discovery bends my model of reality completely out of shape. I tend to spin a worst-case scenario when considering the possibility of sabotage to my cherished world view. Perhaps that is what I am doing now—inventing a picture of reality that resonates with my own, comfortably paranoid, beliefs. I have stretched Syed's discovery to preposterous lengths, taking a single incident and constructing an apocalyptic scenario in which all of human history and personal volition had been hijacked by alien forces. My brand of paranoia has led me to abandon the scientific method and leap to preposterous conclusions. It is time to regroup.

Modern Egypt remains a regional power in North Africa, its major population centers clustered along the Nile Valley and Delta. Once you leave the Nile Valley you are greeted by desert and dunes. Mit Rahina, where the tablet is reported to have been discovered, is located about twenty kilometers south of the present-day capital of Cairo near the ruins of Memphis, the ancient capital of Lower Egypt. Since 1914 there have been at least a dozen significant finds in the area, beginning with the excavation by the University of Pennsylvania of the Temple of Ptah of Merneptah. It stretches credulity to believe that the tablet in Dr. Azad's possession is a recent find. Could it have been discovered much earlier and simply sequestered since antiquity only to be rediscovered anew? If so, why would Dr. Azad suggest otherwise? And why did he mention that the location of the find was in or near Mit Rahina? Whoever discovered the tablet and brought it to Syed's attention was probably fabricating a story to give it an air of credibility. Surely Syed would have grilled the person who brought it to his attention on this matter. I may need to amend my *list*.

As I head back into the house, I receive a text message from Linda: "Was there something more to our conversation the other day than you were letting on?" she texted. The query seems innocent enough, but I suspect Linda is fishing for something more revealing about why I had brought up the subject at the café. Perhaps she senses I might be hiding something. There I go again, manufacturing intrigue without evidence. Perhaps my need for stimulation, something beyond the routine, is driving my suspicion.

Dr. Azad's discovery is plenty of intrigue for now. I need to stay focused. "No, just the usual musings that sometimes pop up on my daily constitution," I lie.

"It is an interesting speculation," she continues. "It got me thinking about the possibility of extracting human history from DNA. Perhaps more is buried in our DNA than meets the eye. What if we could 'reconstitute' a history for every human ancestral line and recapture events from the past, at least in some rudimentary form. You know, going beyond merely reconstructing the physiology and biology of our ancestors to possibly capturing actual experiences from past lives."

After listening to Linda's creative aside it hits me: I've got to get Linda involved! Getting to the bottom of this mystery will require intellect, analytical thinking, and a probing mind—all qualities Linda possesses in abundance. But how can I sell the idea to Syed? I may have casually mentioned Linda to Syed—I don't remember—but I doubt that a conversation between two men about a woman would have involved a detailed list of her academic achievements.

Forget texting, this is too important. I call Linda immediately. "Listen, Linda, what are you doing tonight? I need to talk to you about something very important."

"Well, that sounds ominous. What's going on Jake?"

I have stepped into it now; there is no turning back. I will just have to do my best to convince Syed after the fact. "I can't talk about it over the phone. How about meeting this evening at the Westin in Palo Alto? Say around seven, if that's okay?" I am afraid she will say yes and afraid she will say no.

"Okay, Jake. See you tonight."

Fully understanding the true nature and extent of Syed's discovery will require further investigation and analysis. What are the full implications of the find? What does it all mean? Notwithstanding the ongoing political climate and turmoil in Egypt, it is becoming clear to me that I will need to make plans to visit Syed if I am going to get the answers I seek.

The Foreign & Commonwealth Office has advisories against all travel to the Governorate of North Sinai and limited, "essential," travel to other regions. These restrictions do not, however, include the tourist areas along the Nile River or the Red Sea Resorts. And then there is the standard disclaimer: "There is a threat of kidnapping, particularly in remote desert areas."

Protests and demonstrations are not uncommon across Egypt. The advisory mentions they often happen on Fridays (I'll need to remember that). Of course, they could occur impromptu at any time. Getting caught in "crowd control" could get ugly—I should expect tear gas, birdshot, water cannon, or even live ammunition. Several violent confrontations have occurred recently resulting in many deaths. Cairo and Alexandria seem to be the hot spots and clashes there can be especially violent. The advisory states that, should we become aware of any nearby protests, we should "leave the area immediately." Check. Also, we are advised to "take out comprehensive travel and medical insurance before you travel." Copy that.

The advisory also states there is "serious risk of violence and sexual assault at demonstrations . . . NGOs report several rapes and sexual assaults . . . Foreign and Egyptian women have been attacked."

Well, that certainly puts the kibosh on my plans to include Linda. Or does it? Sure, I should have done further research before even thinking of asking her to come along. And yet, she is certainly no "damsel" in need of my or anyone else's protection. Would it be an insult to exclude her from my plans?

As far as convincing Syed about Linda coming along, that is a wholly different concern. Syed is as progressive as one could hope for given his position and cultural background. He doesn't think twice about calling me by my first name. But still, introducing Linda into the mix could present a new set of issues that I don't feel prepared to deal with. Maybe I won't have to. Either way, I owe it to Linda to at least tell her what is going on. If I do tell her everything, I know she will want to join in on the hunt.

I decide to meet with Linda as planned and explain everything. I will divulge what I know in the strictest confidence and trust her to keep the matter confidential. She might be able to extend spring break and wrangle a few extra days off from university. I will leave the option open. I will also need to explain the many caveats that the FCO had issued, especially as they relate to women traveling in the region. I don't see any other way out of my ethical dilemma.

As I gather the remains of my breakfast (my laptop making a great tray for busing items) and reenter the house, my cell phone rings. It is Syed. I put everything down and answer.

"Syed, do you have something new to report?" I feel anxious hearing from Syed again so soon. My brain has too much to process already.

"I spoke with Khaled Mosi this afternoon and he had some interesting news—"

"Who's Khaled Mosi?" I interrupt.

"He is the gentleman that first brought me the tablet. He tells me there may be additional tablets I might be interested in." My suspicions grow as he continues to explain. "He also told me how he came in possession of the original tablet. When I first asked him about it and how it came into his possession he was very hesitant. I now know why. He says, and this is not surprising, that it was 'acquired' from a previous robbery. I think he had been drinking when he called and blurted it out without realizing what he was saying."

"That explains one thing but leaves others questions unanswered," I say. "The tablet, as well as any others of a similar nature, couldn't have been on display. The importance of such a find would have attracted worldwide attention. Were they being kept in secret for some reason? This makes no sense Syed."

Syed goes on to tell me that during Egypt's plunge into political turmoil looters had taken advantage of the situation and committed many acts of vandalism. The Malawi Museum in the city of Minya about three hundred kilometers from Cairo was robbed a couple of years ago. Perhaps, he offers, the items had been taken from there. As to why or even if they had been intentionally secreted, he cannot say.

I feel uneasy pressing Syed but I have to find out more. Realistically, however, he knows no more than I do about this discovery or why it is only now surfacing. The tablet may have been originally discovered near the town of Mit Rahina or simply materialized from thin air! Perhaps I am melding reality and fiction together to create a fantasy, an adventure designed to distract me from my otherwise routine existence. No matter, if Cypress bought into an artificial reality, so can I. At any rate, my old reality will always be there waiting for me like an old friend . . . won't it?

I might as well throw everything against the wall and see what sticks. "Syed, what would you say to my joining you in our search, in Cairo I mean? Look, I know things are a bit dicey in Egypt right now—"

"What do you mean . . . 'dicey'? Is that an American gambling term?" Syed is still getting up to speed deciphering the meaning behind my occasional use of idioms in my speech.

"No, it just means things are uncertain in Egypt right now with all of the recent turmoil, you know."

"That is quite right Jake, and I wouldn't want to be responsible if anything should happen to you." The concern in his voice is palpable.

"No risk, no reward," I say. "Besides, if what you have uncovered rings true—"

"Ah, I know you are not speaking of a telephone," he says. "It is another of your *Americanisms*. You mean if it turns out to have the 'ring of truth,' if it is genuine."

Syed had learned English when very young and his English-speaking skills have improved considerably throughout his lifetime. "Exactly," I say. "What we need to do now is speak further with this Khaled Mosi fellow and see what else he may know that could further our investigation. What do you say, Syed? Are you game?"

This time Syed knows what I mean without my having to explain. "I will have to rearrange my schedule but I am sure I can free up some time if necessary."

Certain that Syed is committed to the project, I figure it is now time for the *piece de resistance*: "Say, Syed, I have a colleague that might be able to help us. We could use another perspective, don't you think?" I deliberately omit the mention of Linda by name.

"I don't know Jake, 'twos company and three's a crowd.' If things get even more 'dicey' we might want to limit any additional involvement. We could be in danger. You never know."

Now Syed is just showing off his command of the English language. "Well, I'm not even sure if I can get my colleague to agree to come with me," I say, still being evasive about just who I have in mind. Then, without thinking, I blurt it out: "I'd like to give her the option of contributing as I feel she could provide a valuable perspective."

To Syed's credit, he doesn't reject the idea out of hand. He simply restates his concern for our safety and says he trusts my judgment. I guess that settles it then. It's up to me to figure out what to do about inviting Linda. I will know one way or the other by this evening. "I'll get back to you as soon as I have made the arrangements. I think we need to strike while the iron is hot."

"Yes, I agree," says Syed, "there is no need to hesitate. We do not want to be lost." He hadn't exactly nailed the expression but it is close enough. He is willing to move ahead and, as far as my *colleague* is concerned, that decision will be mine alone.

Syed can probably guess that the person I have in mind is Linda, even though he doesn't let on. He may even suspect my motives in wanting to bring her along. I would certainly enjoy her company but there is more to it than that. Linda has a way of quickly getting to the nub of a problem; using her intellect and expertise much as a surgeon would wield a scalpel. I am convinced she will be a welcome addition. If she chooses to contribute anything beyond her many intellectual talents, I will of course be—let us say—amenable.

Linda and I are not what you would call an "item" on campus. We have had dinner together on occasion but it has never gone any further than that between us, not that I would mind. We are just good friends who enjoy each other's company. We especially enjoy discussions of a philosophical or scientific nature. She is well versed in many disciplines, a quick study. At first, I found her ability to grasp difficult subjects uncanny, as I would of anyone not yet versed in a particular specialty. I once casually mentioned a passing interest in neuroplasticity (how the brain can rewire due to new learning or experience) and the following day she gave me a dissertation that contained many insights I had not previously considered. There is no question in my mind that my decision to invite Linda along is sound.

I cannot, however, take my mind off the idea of traveling to Egypt. There is always some element of caution required in any international travel but especially in Egypt due to the shaky political climate. The inherent danger associated with this mission begins to sink in big time and I start to have second thoughts about asking Linda to accompany me.

I decide to relax and do some reading to take my mind off of things for a while. Before I know it, it is mid-afternoon. Time for what I like to call my "power nap." Several famous people have indulged in this practice—Leonardo da Vinci, Eleanor Roosevelt, and John F. Kennedy come to mind, among others. I haven't worked out exactly what I am going to say to Linda. I continue thinking about it for a while after lying down and finally drift off.

I have always had a built-in mental clock that never lets me nap for more than twenty-five minutes or so. I awake refreshed and clear-headed, ready to take on the challenge before me. I forget about rehearsing what I might say to Linda about the discovery and her possible involvement and decide to just let things take their course. I dress business-casual, blue shirt, light grey V-neck pullover sweater, and dark slacks. Good to go.

The Westin Palo Alto is just a few minutes from the faculty housing. It has an excellent outdoor dining area, the "Courtyard of the Moon and Stars," that offers exquisite cuisine while providing ample room to relax and have a quiet chat. In other words, white table linens, straight back chairs, etcetera; all of the accouterments expected of any four-star hotel. A mild evening breeze is coming in off the ocean, a bit cool but tolerable, and perfect for our rendezvous. The lighting is subdued, just enough to hide wrinkles while accentuating one's more attractive features. Linda and I have dined at the Westin before and it seems a good choice. I arrive early and nurse a *Sidecar* (cognac, orange liqueur, and some such) until she arrives.

Spotting my drink Linda says, "I see you are in good company."

"The company just got better," I say, looking up at Linda as she joins me. She is wearing a dark blue dress suit, snuggly fit, with white lace trim. Her dark blue eyes glint as she flashes a friendly smile and joins me.

Neither of us feels like having a full meal. We are alike in this respect; we either skip dinner altogether or eat light. Linda settles on a salad with tea while I have my favorite, clam chowder. I never turn down clam chowder when it is on the menu. The best chowder I ever had was at a little café on the wharf in Santa Barbara. They serve it in a bread bowl so you can devour everything, bowl and all!

Linda has no intention of bantering about pleasantries; she wants to know what is going on. With her elbows on the table, palms up, shoulders slightly raised, she says, "Well don't keep me in suspense! What is this clandestine operation you're working on?"

My mind blanks for a moment. Linda can be distracting. "I don't mean to sound ominous or foreboding—"

"You just did!" she says. "What have you gotten yourself into Jake?" She gives me a look that takes me by surprise.

"Well . . . what I mean is, I don't want to sound deliberately alarmist, you know, like I'm exaggerating or anything." I pause for a moment to let the comment sink in. "Let me just tell you what happened. And, I know, you'll want to jump in right away and you'll have a million questions. But I've got to get through this in one telling or I might miss something and I want to get it right."

Linda raises her eyebrows as she sometimes does when she's getting impatient; but otherwise, she seems content to let me have free reign. "The floor's yours," she says.

I try again: "You remember my mentioning a Dr. Azad? He's the Egyptologist I met a couple of years ago in Paris." Linda is listening carefully, resting the three middle fingers and thumb of her right hand on her cheek. Her only response to hearing about Dr. Azad is to pull her hand away and lift her brow, as if to say, "Yeah, whatever, go on."

"He called me yesterday and told me he had a very old Egyptian tablet. But the thing is, the inscription on it isn't what you might expect—it isn't hieroglyphics, it is an equation. It's known as the Schrödinger Equation and is used in quantum mechanics." Linda's eyes are staring right through me now; she doesn't even blink.

"Understanding the equation isn't the most important thing right now," I continue. "We need to figure out how any information from the twentieth century could end up on a limestone tablet in ancient Egypt. Syed, Dr. Azad, tells me there may be other tablets as well, and, if there are, who knows what they might tell us?"

Linda's attention hasn't waned, a good sign that I am managing the narrative, so far. "I told Syed that I would be willing to join him to help in the investigation. He seemed encouraged but was concerned about my safety. The political climate in Egypt remains unstable. We discussed his concerns and I said I was willing to go ahead and that's pretty much where we left it. He's expecting a call from me to let him know when I will be arriving."

I want to tell Linda about the prospect of her coming to Cairo with me but I decide to wait and see how she feels about everything first. "The situation over there requires caution but I think if I stay with Dr. Azad and don't wander off on my own, I'll be alright." I deliberately downplay the threat to assuage my growing apprehension. Backtracking, I say, "But it will be risky. Any foreigners traveling to Egypt will need to be careful, especially an American."

"I'm going with you!" says Linda. It isn't a suggestion.

"Wait a minute," I say, making it sound like I have to approve. "You can't just drop everything and go; you know the restrictions on faculty leave. You'll need an act of Congress. I'm already on sabbatical and told Syed I was coming. He's expecting to hear from me. I can't tell I'm taking a guest along. And don't take this the wrong way, but if I were to tell Syed that I am taking a woman with me, well, that might be a hard sell." I thought I sounded pretty convincing. If Linda can stretch spring break and Syed is still onboard with her joining us, it might even lift my prestige a notch or two in her eyes.

"Jake, I'm aware of the risks involved; I'm not naïve. But if what you are telling me is true, I don't want to be grinding out lectures and following lesson plans while you're embarking on an adventure! The impact of such a discovery would be enormous! I can't sit around waiting for you to text me with updates—I want to be a part of it! This Dr. Syed or—what is it, Azad—you've got to tell him how important my contribution could be. Jake, I can't miss out on this opportunity!"

Linda is practically pleading; she sounds desperate. I have to let her off the hook. This just isn't right. "Okay," I say, "I'll talk to Syed. I'm sure I can convince him. But you'll still need to arrange some time off. And remember: We don't want anyone to know about this until we have nailed everything down solid. The last thing our careers need right now is for both of us to end up a laughing stock, not to mention the damage that would be done to the university's reputation."

I hand Linda a copy of the list I've prepared detailing some items I want to follow up on. "I'm sure you can add to it."

"Well, obviously authentication would be a prime concern," says Linda. She seems momentarily distracted as the import of what we are dealing with catches up with her. "I'm sure Dr. Azad is more than familiar with the process. Tell me something, how do you tell if an object—even if it has been authenticated as being very old—hasn't been tampered with more recently? You could look at the molecular structure I suppose, the way I do in my work sometimes. But dealing with non-organic material . . ."

This is exactly the type of forensic question I have come to expect from Linda; another reason for including her in the investigation. "That's a good point," I say. "They probably have chemical assays and other techniques to determine if there has been any tampering. I'm not sure. Maybe Syed could enlighten us."

Getting back to the matter at hand, I press her further on the possibility of requesting some time off to join me. "I think I'll be alright," she says. "I don't know, we'll see. I'll let you know as soon as I can." She doesn't seem too worried about it so I let it go.

"So, this is what prompted your 'scenario' the other day," says Linda. "What do you think it means?"

I pause a moment to let the attendant place the salad, tea, and clam chowder on the table. "Unless we're the biggest dupes this side of the Atlantic," I say, "someone or something has either sent a message backward in time or ancient Egypt had knowledge of modern science. I don't know how to make any sense of it. You're the one with an analytical mind; I'm just the mechanic. What do you think's going on?" I have several working hypotheses but want to get Linda's input before commenting.

"I would begin with the null hypothesis that the plate, or plates, have been misinterpreted or simply misunderstood. Someone placed the equation on a tablet or slab using materials from ancient Egypt. They're fakes, forgeries. As to why your Dr. Azad has not been able to figure this out, I couldn't say. But as interested as I am in wanting to believe something mysterious is going on—and I'm still on board with this whole thing—you must admit the possibility of the find being genuine is remote. I don't see a way, barring some type of time travel or interdimensional teleportation that these tablets could have shown up where they did. If ancient Egyptians had such knowledge there surely would be other evidence, abundant evidence."

As usual, Linda has zeroed in on the most likely possibility. I don't want to consider the tablet a fake and spoil my chance of participating in no small adventure, but Linda is probably right. But still . . . there's a chance she could be wrong and I refuse to accept her suspicions out of hand. "But what if it isn't a hoax," I say, trying to salvage a chance to keep the possibility alive. "Syed certainly knows what he's doing and he wouldn't have called me if it was merely a hunch."

Linda has been nibbling on her salad, looking up occasionally to let me know she is listening. She then suddenly launches into a playful mood that seemed so completely out of character from my first impressions of her. I always thought of her as being so serious, the only adult in the room, until I got to know her better. She can laugh and joke and banter with the best of us when she wants to. And when she does, you just have to laugh with her; her lively, animated, engagement draws you in. She often reminds me of the physicist Richard Feynman. There is a photograph of him interacting with his students that shows his face lit up in wonder and fascination. She has that same look now. "So, what's your take on all of this?" she says, her expression vibrant, eyes open wide. "You're the physicist. What are we dealing with, wormholes? Cosmic time travel? Flux capacitors?"

I laugh and then she laughs and then we both crack up. It is at times like this I feel especially drawn to her. I don't exactly know how old she is, maybe in her late thirties. I sometimes think there might be too much of an age difference between us but maybe that's just me. She might not feel that way. Recovering my composure, I decided to present my alternative hypotheses.

"Backward communication in time has not been ruled out," I begin. "If time and space are similar in some way, it becomes possible for actions generated in the future to affect the past. My current work in quantum mechanics is related to this idea; specifically, a concept called entanglement. The upshot of entanglement is simply that two particles produced together, when separated, remain in a fixed relationship to each other. Inducing any subsequent change in the condition of one particle is instantly matched by a corresponding change in the other particle. What is so startling about this phenomenon is the instantaneous nature of the corresponding alterations taking place. As confusing as it may seem, this choreographed dance does not require a faster-than-light, or superluminal signal, to pass between the entangled pair." I pause for a moment to make sure Linda is following me before continuing.

"In my research, I came across a paper published by George Musser who used to be a senior editor at Scientific American. He cited Huw Price; a University of Cambridge philosopher involved in the physics of time. Musser seems to be saying that, according to Price, interaction with either entangled particle causes effects to travel backward in time to when the particles were originally closer together and strongly interacting." To make sure I don't misquote the author's astounding conclusion, I pull out a folded sheet of paper, a copy of the article containing the key point I want to emphasize, and read verbatim: " 'At that point, information from the future is exchanged, each particle alters the behavior of its partner, and these effects then carry forward into the future again.' "

Anticipating an objection, I continue: "Now, before you tell me that Egyptian tablets from several millennia ago are not 'particles' and couldn't possibly become entangled across such a vast expanse of time, consider the possibility that sometime in the future—who knows how far—it becomes possible to quantum entangle entire objects, not just particles. As far as the time barrier is concerned, perhaps at some deeper level, we don't currently understand, all perspectives—past, present, and future—are the same. I mean, who knows? Sure, it's conjecture, but if what Dr. Azad has uncovered is real, we can't rule it out entirely. Maybe the entire universe is already entangled in some way that we have not yet discovered. After all, wasn't everything 'entangled' at the beginning?"

Linda's next question assures me she understands what I am suggesting. "So how do you scale up from the quantum to larger objects? And how would a plate from the future correlate or 'entangle' with one from the ancient past? Are they both somehow together in the future and—"

I jump in before she can complete her thought process. "It may just be that a signal or image was sent backward in time and was somehow inscribed on the older tablet. Or, if the two tablets—one from ancient Egypt and one from the future—are somehow truly entangled, inscribing or otherwise affecting one would affect the other. Why or especially how an inscription would translate identically across space-time without a correlated 'distortion' I couldn't say. None of this makes much sense; if it did, I wouldn't need to construct these hypotheticals in the first place. I've also considered another version of entanglement called entanglement swapping that is even more bizarre but I'm holding off on this idea for the time being. I haven't even ruled out human time travel but I'm not prepared to go there unless that is where the evidence unmistakably takes us. Either way, the truth might be more unbelievable than we can imagine."

I decide to forgo further speculations. No sense adding to the confusion. Better to leave tesseracts, higher-dimensions, branes, gravitational anomalies, worm-and-black holes in my quiver unless and until needed. But nothing is ruled out as far as I am concerned until I know for sure what is going on.

"You said there might be other tablets," Linda probes. "They could be the other pieces of the puzzle we need to make sense of everything, tie it all together." Linda is right . . . again. In all of the excitement, we both lose our appetite and decide to call it a night. We agree that we first need to gather all of the information we can before we have any hope of tracking this mystery. As Sherlock Holmes would say, "The game is afoot!"

Chapter Three: The Journey Begins

On my drive home from my meeting with Linda, the thought of traveling to Cairo again crosses my mind. Tensions in the Middle East are still ripe and I am concerned on two fronts—our safety once we arrive in Egypt and our ability to conduct our investigation without government interference. The Islamic militant group known as ISIS or the Islamic State continues to vie for prominence in Northern Africa while issuing threats to Europe. Their closest encampment in Libya is just 500 miles from Italy across the Mediterranean Sea. The political turmoil shows some signs of easing but I am still worried about travel advisories to the region.

    The bulletins coming out of the Foreign & Commonwealth Office regarding travel to Egypt are constantly being revised and updated. The caveats fall into three categories, broken down by region: "Advise against all travel," "all but essential," and, "See our travel advice before traveling." The usual caveat is added to the FCO website to cover all bases: "There is a high threat from terrorism." The extreme Northeast province carries the gravest warning but I can't see any reason we would need to visit this area. Egypt, predominantly an Islamic country (The official name being the Arab Republic of Egypt), requires any visitors to respect local traditions, laws, and customs. We would need to exercise caution as not to offend religious beliefs. The holy month of Ramadan would not start until June so those observances would not be impacting our visit. Still, what might be considered normal in America, or at least not risqué, could bring approbation if not legal sanctions in Egypt.

There are additional admonitions regarding dress: "Women's clothes should cover the legs and upper arms. Men should cover their chests." Public displays of affection are "frowned upon." Drinking alcohol anywhere in public other than a licensed restaurant or bar "is not allowed and can lead to arrest." I don't take the fact that these warnings are "advisory" too lightly. The Egyptian authorities, no doubt, would take them quite seriously.

Okay, no problem. We can certainly adapt to local customs and ordinances. The real concern, the penumbra under which we will operate, is our safety. Circumstances are fluid and a violent disruption could trigger intervention by civilian and military authorities which, in turn, could lead to even greater danger. I am determined to press forward come what may. Perhaps I am being foolhardy, but my resolve remains unshakeable; I have to unlock the hidden key to this mystery.

I have traveled to Europe several times and have a current passport. I know Linda had traveled outside the U.S. as well. She told me she had spent some time in England after completing her postdoc in molecular biology and genetics from the University of California at San Francisco. That was some time ago. If she needs to acquire new documents, that might slow things down a bit. One more wrinkle.

The U.S. Department of State website, under "U.S. Passports & International Travel," does not list any restrictions regarding currency whether you are coming from or going to Egypt. There are no vaccination requirements which seems a bit odd, especially now. A tourist visa is required but, according to the official site, one can be obtained easily enough at the airport in Egypt. Despite travel advisories that might suggest otherwise, getting to Egypt doesn't seem to be a major hurdle. Tourism continues to play a large role in boosting the Egyptian economy despite the "Arab Spring" and other disturbances the country has experienced in the past. The current information available shows the contribution of tourism to Egypt's overall GDP exceeding eleven percent. When it comes to properly assessing the situation, I will be relying on Dr. Azad. He will have a better understanding of our prospects of arriving and completing our investigation successfully.

Most of the flights from San Francisco to Egypt are routed through Germany, either Frankfurt or Munich. Lufthansa, United Airlines and Egypt Air flights, in various combinations, complete the round trip. Before booking a flight, however, I need to know if Linda will be able to wrangle some time off to join me. The March spring break will allow some time during mid-month. That might have to be enough. The last day of classes is Friday the thirteenth. Even though I am not normally superstitious, the thought makes me a little uneasy. The only other items I can think of are Linda's passport status and last-minute details Syed might want to throw into the mix.

It is now late Monday evening. If I leave for Egypt Wednesday morning, I should arrive by late Thursday afternoon or early evening at Cairo time. That might be jumping the gun a bit but I am anxious to arrive before Syed follows up with his original contact, Khaled Mosi. Linda should be able to book a flight Friday if her schedule permits and join me in Cairo. Am I being selfish? Should I at least wait for her to get back to me before making any plans? Probably. But I'm too excited to wait. I book an outbound flight leaving Wednesday morning, fortunate to find a seat under such short notice.

Let's see, nine hours, nine p. m. here Monday, six a. m. Cairo time, tomorrow morning. I'll get the hang of it. I'll call Syed and confirm my itinerary and then—Wait . . . I don't even know if Syed can free up his schedule to accommodate my visit. If there was a problem, he would have already told me. No matter, I'll call and see what's going on. I often keep a running monologue going in my head to help me keep track of things, occasionally extending it to a full-fledged dialogue that includes two versions of me bantering back and forth, sometimes takings sides for or against my internal argument. Doesn't everybody?

I call Syed and reach him at home. Before I can say a word, he quickly offers his patented greeting— "Professor Jake, good to hear from you again so soon." Syed sometimes calls me "Professor Jake" instead of Professor Banner when in a spirited mood; it is part of his charming personality. "I hope you have good news for me about your trip," he says. "When should I be expecting you?"

"I have everything arranged. I should be arriving at Cairo airport shortly before five p. m. Thursday."

"Most excellent," says Syed. "I shall be there to receive you. We will have some dinner and speak more about this mystery. I also have some more good news for you. And I am anxious for you to see the item we have been discussing. It has created a great"—he pauses for a moment to gather just the right word— "conundrum. Is that the right word?"

"It is the perfect word, Syed, and we will unravel this 'conundrum' together."

Syed is being slowly pulled into the hunt. "I am encouraged to have you working with me Jake, but what about your colleague? Will I need to prepare another room?"

"Well, I don't know just yet. She should let me know in the next day or two. What rooms are you talking about? I have already made arrangements for my stay." I had no idea that Syed has prepared to open up his home for me.

"Nonsense, you will stay with me. I have plenty of room. I am sorry you have made arrangements. I should have let you know I would be receiving you. I hope you are not holding a fee for your room." Syed is genuinely sincere in his offer and his tone reveals his concern. "Where have you made the arrangements?" he asks.

"Well, let me see; just a minute." I grab a copy of my itinerary. "It is the Hilton Zamalek in Cairo. I'm sure it will be fine—"

"Just hold on a moment, Jake. That would be Zamalek Island, yes. Hold on."

Syed puts me on hold for some time and, when he returns, he says, "It is all okay now. Your reservations are canceled. There will be no problem."

My first reflex is to ask Syed how he managed to cancel the reservations and fees with one phone call but I let it go. I am beginning to suspect there is more to Dr. Azad than meets the eye. At the Paris conference, I recalled several guests coming over to greet Syed. I had noticed deference in their demeanor when speaking with him and they hadn't seemed like the typical academics I'm used to. "That's wonderful Syed but I don't want to impose—"

36

"No more about this Jake; it is done. I will be looking forward to seeing you. You will have the layover in Germany of course. Will you be stopping over in Frankfurt or Munich?" I had no idea how often Syed entertains visitors from across the Atlantic but it is nevertheless impressive that he knows the flight patterns of foreign travel from America, at least those traveling business class. Depending upon my choice of airline, the layover could be different or even a direct flight. Maybe he is just guessing.

I refer again to my itinerary. "I should be in Frankfurt Thursday morning."

"Then you will call me during your layover and let me know of your progress. I will be expecting your call. It will be good for us to see each other again."

"I will call you then," I say. "See you soon." And that was that. Everything is set. All I need to do now is get my things together and . . . and what? I don't know. It will all get sorted out one way or another. But I still need to call Linda tomorrow to see if she is making any headway with the administration. The strictures at Stanford regarding faculty leave are formidable, but she may be able to work something out, you never know.

Syed is just a professor—well, maybe not "just" a professor; after all, I'm not "just" a professor, but he isn't a diplomat or statesman. Does he have connections, political connections that I don't— or maybe shouldn't— know about? Or am I, once again, reading too much into things? Maybe—molehills sometimes appear larger than reality would suggest. However, the images on the outside mirror of your typical sedan are closer than they appear. I put the matter on my mental shelf, subject to further review.

I have flown enough times to learn that the lighter my luggage the better, within reason. Sundry items can always be procured at my destination if I forget anything trivial. I will bring a semi-formal suit, just in case. Other than that, casual should do. Anyway, it's getting late. I can sort out any remaining details tomorrow.

I wake even earlier than usual, anxious to get underway. After completing my packing, I decided to have a light brunch. I should still have enough time to review a couple of scientific papers that might prove useful. One of the papers deals with distortions in the organization of matter at the atomic scale and how these alterations relate to the chemistry of the material under study. The other paper suggests, and attempts to prove, that time in the quantum world is demonstrably different that in the classical world. It submits that time runs both forward and backward. Although these ideas do not in themselves prove anything immediately useful, they might point in the right direction. If these areas of research prove prescient and are realized in a much larger context in the future, they could hold the keys to our puzzle . . . or not. All options, possibilities, remain open.

The kernel of these ideas, as far as I can see any possible connection or relevance to our investigation, is the very nature of time itself. The essential question becomes: Can events that will be taking place in the future affect events that have already taken place in the past? The normal way of looking at time demands we dismiss such a notion out of hand. And yet, here we are. As to how things will ultimately unfold—the denouement—my only thought is: Time will tell.

I jump into my BMW convertible and head out for The Annex at St. Michael's Alley in Palo Alto for an early brunch. I have always permitted one indulgence—the best luxury vehicle I can afford. My bimmer fits the bill. "Afford" is a bit of a stretch but that's the point. I rationalized the decision, believing everyone has their druthers when it comes to extravagant and "unnecessary" purchases.

The Annex is somewhat cramped for room and there is often a line, but it is worth the wait. It has what some might call a yuppie or even snooty atmosphere, but I never go there for the ambiance. I order my usual French toast with pan de mie bread and powdered sugar, sans strawberries. It is another of my indulgences but I feel I have earned it after my morning workout. I keep my weight between one hundred sixty and sixty-five pounds. I never bought into the myth that a man's weight is a measure of his masculinity. Oliver Wendell Holmes, Jr., former Associate Justice of the Supreme Court, stood six-foot-three inches tall and, in his youth, weighed one hundred and thirty-six pounds. My weight is just fine; maybe I should drop a few pounds!

After brunch, I head back to what I like to call "The ranch." My home is anything but a ranch but it is a favorite expression of mine, as in "See you back at the ranch." The furnishings are modest, eclectic, a smattering of paintings discreetly positioned throughout. I would prefer original paintings but have settled for "artist's proofs" and sundry prints. I call my favorite room "The book room." Not original by any means but it captures the essence—books. I read that shortly after Johannes Gutenberg introduced printing to Europe in the fifteenth century the equivalent price for a modern book today versus then would amount to twenty thousand dollars. By these standards, my "library" would be worth a mall fortune.

I cannot *not* read. In my formative years when I was exploring philosophy, I ventured into the works of Plato, Socrates, Balzac, Shakespeare, Spinoza, Kant—the usual litany, among many others. Even today I can usually be found perusing the latest thinking of contemporaries expounding on a variety of subjects: law, medicine (neurology in particular), philosophy; and especially physics. Most people I know suspect my political persuasion to be muddled and confused. I can as easily be found reading Alan Dershowitz or Jeffrey Toobin in law as Mark Levin or Ann Coulter, either of whom would make a staunch liberal or conservative cringe. I prefer to think I am informed rather than confused.

My greatest enemy is certitude—not so much in matters of faith or personal conviction as in matters of science and our general inquiry into the nature of reality—legal, political, philosophical, or otherwise. My model of reality is not sacrosanct, inviolable; it is as subject to review and revision—based upon new evidence or a fresh perspective—as another's. I find intolerance intolerable. I welcome disagreement, even heated discussions reflecting differing viewpoints. It is dogma and groupthink I despise. If the only reprobation I receive for an open mind is to be thought confused, I can live with that.

I am completing some last-minute details for the trip when my cell phone rings the familiar ringtone from the TV series "24" where Kiefer Sutherland played the Counter Terrorist Unit (CTU) agent Jack Bauer. I'm no Jack Bauer but at least we have the same initials.

It is Linda calling. Hoping to hear good news I answer in a positive tone. "Hey," I say, expecting to hear Linda's usual upbeat reply. "What's up? What's the good news? Are we good to go?"

"Hey, Jake," she says, in a noticeably somber tone. "I'm afraid I have bad news. The earliest I can get away will be early afternoon Friday. I just have too much on my plate so close to spring break. I booked a flight as soon as I could."

I sense the frustration in her tone; she is despondent. "I spoke with Syed and he said he would be able to accommodate our visit. He said he has the room. It didn't seem right to refuse so I thanked him and told him I would see him as soon as I could. Will you be able to stay beyond break if we need more time?"

"It doesn't look like it," she says, trying to sound like there might be a slight chance she can gain some wiggle room.

"It's all good Linda," I say like it's no big deal. "I'll just set up shop and we can nail everything down when you get here. Oh, by the way, were you able to get your passport in order?"

"I'm all set, Jake. Don't worry about anything. I'll call you as soon as I can. I gotta get back."

She seems to be holding up under the circumstances. No sense pushing it. "Okay, talk to you soon."

I feel bad about the situation. If Linda and I could travel together to Egypt together we would have more time to map things out and better plan our strategy. But, as it is not to be, I let the matter go. The rest of the day is somewhat of a daze. My thoughts alternate between delusions of grandeur where, on the one hand, I become world-famous for deciphering an impenetrable enigma and, on the other, a laughing stock for participating in an elaborate hoax designed to deliberately mislead the scientific community. I suspect I won't be sleeping too well tonight.

Chapter Four: Tangled Web

After clearing international travel check-in procedures, I board my flight for Frankfurt. I was fortunate to get a window seat when making reservations; it will be a long flight.

I have learned from personal experience while traveling to keep my profession nondescript; otherwise, there would be no end to the relentless questions I would be subjected to—especially if my seat-partner fancied him-or-herself to be an "expert" in my profession. The gentleman sitting next to me seems safe enough—a middle-aged man with a slightly receding grayish-white hairline, heavyset, with a friendly enough demeanor. He sits down, smiles, and gives me a polite greeting: "Hi, my name is Tom; well actually Thomas, but I prefer Tom. How are you?"

For some reason, as often happens when I am feeling uncomfortable, I entertain the most inappropriate thoughts. I am reminded of a classic movie, the name of which I do not remember, where one gentleman, being asked how he was doing, replied: "Well, it's pretty much like the man falling from a tall building. Every floor he could be heard to say, 'So far so good . . . so far so good.' "

"So far so good," I say, hoping that will suffice.

"So, off on another adventure?" he says, trying to get a conversation underway.

Even though I hadn't slept too well last night, it is still early. I don't feel too groggy and it will be a long flight, so I take the bait. "Oh, just off for a couple of weeks or so, nothing special, How about you?"

"I'm going to see my daughter. She lives in Cairo; married a fellow named Tony who works for Barclay bank. She met him in New York where she lived before they got married. He had returned to the U.S. on business when they first met. It must have been a whirlwind courtship. She called me after they had been seeing each other for only a little over a month and told me she was getting married. I was a bit stunned but I guess these things do happen. My wife and I courted for over two years before we married. He's an American, born and raised. Pretty remarkable guy, speaks five languages, and heads up the Cairo Branch. Doing pretty good from what I can tell. I'm looking forward to the visit; should be interesting. It's my first time to Egypt, how about you?"

"Same," I say, recovering from the drawn-out monologue. I begin to think I will escape the usual lecture on my many failings in properly understanding my profession when Tom says, "So, what do you do for a living?"

I think quickly. I have three options—dodge and weave (that never works out too well; I usually get pinned down sooner or later), describe what I do in vague enough terms that I might just get by with a brief explanation (sometimes works), or just let it out and hope Tom is sufficiently ignorant on matters of physics to drop it altogether and move on to something else.

We never really know what we are going to do or say from one moment to the next. Dr. Sam Harris, who has written several books on morality, ethics, and free will believes free will is an illusion. We cannot decide what we will next think or intend "until a thought or intention arises (subconsciously)," he tells us. He then sums up by asking himself: "What will my next mental state be?" His answer: "I don't know—it just happens. Where is the freedom in that?"

Not knowing what *I* will say next, I blurt out "I teach science." Maybe if I keep it general, he will lose the scent and call off the hunt.

"Oh yeah, what's your subject?" he says, probing further.

"Physics" I respond. There, it's out in the open. I brace for his response.

"I never did too well in science; it wasn't my best subject."

And that was that. I escape unscathed. The rest of the flight to Frankfurt goes smoothly. Tom and I enjoy friendly chitchat about our favorite movies and other noncontroversial subjects. I am careful not to bring up anything religiopolitical.

I don't ask Tom about his wife, why she is not with him. It's his business. If he didn't bring it up why should I? And then, as if reading my mind, he tells me about Marsha. They had been married twenty-six years before she died of a rare illness a few years ago. She had what he called Brugada syndrome which causes a disruption of the heart's normal rhythm that can often, as in her case, lead to death. I express my sympathies; I don't know what to say. I do, however, know what *not* to say—something dumb like, "I know how you must feel."

As Tom recounts his wife's death it gets me thinking again about Dr. Azad's discovery. I wonder if any of the other plates if they exist and we can track them down, hold out any promise. Maybe they contain medical advances—breakthroughs in medical technology or innovations in diagnostics and treatment. Or are they all confined to mathematics and physics? I am anxious to find out.

I excuse myself and shut down for a long and much-needed rest, out like a light.

The familiar sound flyers all over the world love to hear wakes me from a deep sleep. I open my eyes to get my bearings. Just turbulence; we still have a way to go before we land in Frankfurt. Funny how pilots like to wake passengers up to announce turbulence, like nurses in a hospital waking up patients to make sure they took their medications, often prescribed to help them sleep.

Tom is unfazed, sleeping like a baby. I grab a paperback book I had purchased to read on the flight, an FBI agent's account of his time in the bureau. It lays out his insights on crime and the future of surveillance. Barely twenty pages in I begin to regret my choice. As I continue reading, I become increasingly convinced of the utter hopelessness of securing data. The author tells us that we are the product of Facebook and Google and so many other social websites — unpaid dupes that willingly hand over all of our data to them so they can, in turn, sell it to others for profit. No wonder these "services" are free! The companies that purchase our personal information even use our images and personal information to promote their products. It reminds me of the saying that, when you find yourself in a "friendly" card game with a group of strangers and you are wondering who the "dupe" is, it's probably you! The author promises to provide remedies for my mounting fears if I can manage to plow through to the end. I might as well forge on. I hadn't brought anything else to read and I have to read something. My only other option is—God forbid! —the in-flight magazines.

After about eleven hours, give-or-take by my count, we begin our descent. Hungry as I am, I decided to save my appetite and forego airline food, settling for snacks and bottled water during the flight. I will have something to eat after deplaning in Frankfurt. Airport food won't be much of an improvement but I don't want to deal with the hassle of exit and re-entry to and from the terminal. Besides, it is more than a bit nippy out. I brought a light jacket but suffer from what I call "pansyitis"—a condition wherein one is unable to tolerate extreme variations in temperature. Anyone born and raised in California or who has lived there for any appreciable time develops the malady and I am no exception. I could rough it if circumstances demand; but otherwise, not.

I had left the West Coast about eleven a. m. Wednesday and it is now close to six a.m. Frankfurt time Thursday. Let's see, Frankfurt is one hour earlier than Cairo, so that would put it at just about seven a. m. in Egypt. I can expect good weather once I arrive. The temperature varies considerably between Germany and Egypt. It should be much warmer.

I feel sorry for anyone who is forced to spend more than an hour or so at any airport, international or domestic. At least I had a book to keep me company. The airport is modern, contemporary, clean; comparable to any American counterpart. I had slipped away from Tom when we deplaned. He's a great guy but I didn't want to have to entertain him for five hours.

My internal clock informs me it is past cocktail hour but it is out of sync with Frankfurt time. The drunk's refrain— "It's five o'clock somewhere"—comes to mind. I settle for breakfast and coffee. I then call Syed to let him know there have been no hitches so far, give him my flight number and expected arrival time, and, after several more hours of downtime, head for the last leg of my flight.

Despite pollutants in the air—far too common—the Giza pyramids are still visible, though distant, and in view as the plane descends into the Cairo Airport. It is late afternoon by the time I arrive. Surprisingly, there are no surprises as I navigate through security on my way to meet Syed. When I first see him, he is standing near a kiosk speaking to another gentleman. They are carrying on a conversation in French and he doesn't notice me as I approach. I know Syed speaks French and English in addition to Egyptian Arabic. At the conference in Paris where we first met Syed spoke English but occasionally switched to Arabic when conferring with his colleagues.

After clearing customs, luggage in hand, I head toward the receiving area. Syed turns his head and, upon seeing me, cracks a wide grin, hugs me (I remember the advisory warning against such displays), and says, "Jake, you are finally here. How was your flight?" The gentleman he had been speaking with looks on in silence, an aloof expression on his face.

"Glad to be here . . . I'm exhausted," I tell him, happy to get the last leg of the flight behind me. Jet lag is kicking in and I will need some time to recoup; but otherwise, I am looking forward to a lively interchange with Syed over his recent discovery.

Syed introduces Emile as "our" driver. Emile is about Syed's height, close to six-foot, his hair a mixture of dark black and charcoal gray. He looks about my age, in his late fifties, slim built but sturdy frame. Making every attempt to sound friendly, Emile says, "Messier Banner, Welcome to Cairo. How are you, sir?" His French accent is thick, almost affected. He pronounces my last name "Van air."

"Good to meet you," I say, looking to Syed for a sign that we can get underway.

Syed, sensing my eagerness to get moving, looks at Emile and says, "Help Dr. Banner with his luggage. Quickly, we must be on our way."

Emile opens the car door for us and secures my luggage. Syed and I sit down and he immediately pushes a button raising a glass partition behind Emile. We head down Airport Road toward what Syed calls Hyde Park in "New Cairo," a relatively short drive from the Cairo airport.

During the ride, we avoid what we both come to call "the discovery." I ask Syed the typical questions expected of tourists: "What is the significance of the hijab?" "Why do some of the men wear such long clothing?" Syed is very polite in answering these and other naïve questions I pose.

Turning to the topic of interest Syed anticipates my next question, speaking at length about various dating methods: Radioactive carbon dating; potassium-argon; fission-track dating using energy deposition solids; thermal and optical luminescence. I am familiar with carbon dating. Why he mentioned other methodologies I have no clue; showing off again? He also speaks at length about stratigraphy, a branch of geology that studies rock layers and layering. Any doubts I had about the authentication process Syed used in identifying the tablet quickly vanish. Syed knows his stuff.

The entrance to the private gated community where Syed lives is just off of Ring Road. We clear the controlled entry point and head toward his residence. I notice several security vehicles along the route. Hyde Park, along with "Beverly Hills" and "Dreamland," is one of several "satellite cities," where the well to do live. The names are a reflection of the continuing Western influence in "New Cairo."

Once we arrive, Emile jumps out and takes my luggage. My first thought is, how does a professor afford a home like this? I deserve a raise! The word villa comes to mind. The outside could best be described as pinkish-beige if there is such a color. Entering into the foyer, I imagine how easily it might accommodate a typical Egyptian family with room to spare. I had read that there are hundreds of slums around Cairo.

Syed gives me a tour which is quite impressive: four bedrooms, five baths, a game room sporting a billiard table, two separate terraces—one in front and another overlooking the courtyard in the back—a garden (I don't imagine the water is easy to come by) and servant quarters downstairs in the southwest corner which remain empty. All the comforts of home.

Syed shows me to my room upstairs as Emile follows close behind with my luggage. The room is inviting, lightly decorated, with a large bed in the middle of the room. There is a window just to the right of the bed and an archway next to it that leads out onto a private balcony. There are two separate murals on the walls, both similar, depicting scenes of the Nile with pyramids in the background. What looks like a wide bookcase or storage unit with open shelves stands at the foot of the bed. There is also an en-suite bathroom. Another *Americanization* I suppose.

"Feel free to freshen up and get some rest," says Syed. "You will need to adjust your clock."

"Thanks. I think I'll do that," I say, grateful that Syed is letting me off the hook, I could use another one of my "power naps." It has been almost a day West Coast time since I boarded my flight in San Francisco and I am spent. I check out the bathroom facilities and—lo and behold! —a shower! Maybe it is just this one room that has one; it doesn't matter. I shower and put on a robe hanging on a hook in the bathroom. I sit down on the edge of the bed for a moment and then lay down to rest my eyes. My internal nap clock goes offline and I sleep until almost nine o'clock in the evening Cairo time.

I wake up refreshed and ready to take on the day, or in this case, night. I change, brush up a bit, and head downstairs. The main room downstairs is quite large and the furniture is generously spaced throughout. Three sofas, two stylish looking chairs, and assorted tables complete the arrangement. The center table is large, oval, with an intricate design carved into the surface which I do not recognize.

Syed is sitting on a sofa in front of a large window that stretches upward and is level with the second-floor landing. The curtains are open and the outdoor lighting allows a wide view of the landscape as I descend the staircase—Sycamore and mulberry trees, other varieties I don't recognize, and a beautiful garden sprinkled generously with Egyptian Lotus flowers. A small pond with Water Lilies floating gently on the surface is also just visible.

Syed is absorbed perusing some papers but looks up when he sees me. "Jake," he says, "I thought you were—how do you say—'down for the count.' Come and join me."

"You never get used to jet lag," I say, yawning. "I'm sorry I couldn't keep you company."

"No, no, not at all; come and sit down." Syed sounds upbeat and energized. Perhaps he had taken a power nap.

I sit in a chair next to Syed. "It feels good to relax," I say.

"I shouldn't think you would be very comfortable in that chair," he says. "It is a Timothy De Clue design which I purchased mainly for show and to complement the décor. Most of my guests prefer to sit elsewhere."

It just looks like a curved wing chair to me. Odd that he would remember the designer's name. Is he being pretentious? I am beginning to suspect Syed is very jealous of his reputation and status, not only in the local community and academic circles but throughout Egypt and the Middle East.

"So, what is your schedule like this week?"

"I am free for as long as needed," he replies, looking at me in a way that suggests the question is unnecessary. "I will show you the tablet in the morning." He then changes the subject, asking if I want something to eat.

"Oh, I'll be alright," I say. I don't want to impose as it is already late.

"Would you like some fruit, at least? We have some orange and watermelon slices which are quite tasty. I will get some for you."

"If it is not a problem," I say. I could use something in my stomach besides airplane snacks and airport food.

Syed leaves for a moment and returns with a generous assortment of orange and watermelon slices. As he hands me the plate, he asks: "So what are you thinking of my country, Jake? I would like to hear your first impressions."

I hesitate for a moment, not wanting to say anything that might sound too judgmental. "It is very different from Europe or especially America. I would like to know the customs better and the people—"

"There is much that is misunderstood about my country Jake. Don't believe all you hear in the news. We are fighting tyranny even more so than in the West. We are buying ordnance from the French; with their loans, of course. We have sent bombing raids on ISIL. Our courts have declared Hamas a terrorist organization and we continue . . . Well, now you have got me started." Syed catches himself. "Let's have a drink. I have a good Stella Lager I think you will like." Not bothering to wait for my reply, he takes off to fetch our drinks.

Syed, although a Sunni Muslim, does not engage in moral arguments regarding the use of alcohol. As far as he is concerned, if he drinks in moderation and uses discretion, it is between him and Allah and no one else. Ramadan, observed during the ninth month of the Islamic calendar, beginning in June, is an exception. Not even Syed, with his relatively liberal attitude according to Western standards, would dare stray from strict observance.

Syed returns with an opened bottle of beer, no glass. "Tell me, Jake," he says, "what is your current area of research? What are you working on?"

I take his queue and follow the discussion to a safer harbor. "My basic area of research concerns the relationship between quantum mechanics and general relativity. The subject is well-trodden but it will have to bear fruit eventually; otherwise, we will have to abandon the effort altogether and come up with a completely different paradigm."

What am I doing? "Well-trodden? Paradigm?" Syed's English is good but I can do a better job of explaining; let me try again. "Gravity holds things together on a large scale. On a smaller scale—what we call the quantum level—things are governed by a different set of laws. These two realms, or environments, behave in ways that are not compatible—what happens in the quantum world, the world of the infinitely small, cannot be scaled up to the larger one where gravity governs everything. It is as if two explanations work perfectly well for each of these worlds, but each explanation contradicts the other. One view is based on classical physics and the other on probability."

"So, this theory or idea of yours would join the two worlds together? What would you then call it, quantum gravity?" Syed either knows more about these subjects than I would have thought or he is just guessing; either way, it is impressive that he came up with this formulation or term so quickly. He isn't questioning what I was saying; he understands.

"Very good, Syed, I will make a physicist of you yet! You are close to the mark. It is just such a theory that would help us integrate these concepts into a grand unification theory, what we like to call a 'Theory of Everything' or 'TOE.' "

Syed laughs. "It would be a very big 'TOE' would it not?"

I laugh as well and, reciprocating, ask him about his work at the University of Cairo.

"I do not spend so much time at university any more. I enjoy giving lectures but pretty much keep my schedule. I lend my name and prestige to the university and, in return, they allow me a good deal of freedom to come and go as I please, as long as I complete my lectures."

"If you don't mind my asking," I say, "how do you manage to own such a nice property?" I hope I had put the question properly and didn't appear too nosey, which of course I am. I am curious how Syed can afford so much on a professor's salary. I then have second thoughts about asking such a personal question. "I'm sorry, Syed, it's none of my business, I shouldn't have asked."

"That is quite all right. It is a reasonable question to ask. I have been very fortunate. I have made many good investments, many in the futures market. If you have the right information at the right time, you can do rather well. Take the oil futures, for example. The cartel used to dictate oil prices, and they still have some influence, but the fracking industry changed all that. I was very fortunate. I retired from active investing in oil futures just before market forces tipped the scales in unpredictable ways."

"I wish I had known you earlier," I say. "I might have retired by now."

Syed's explanation about how he acquired his wealth seems simple enough, maybe too simple. There I go again, creating intrigue where none exists. My purpose in being here in the first place involves enough intrigue without having to manufacture conspiracies. There is no need to read anything more into what Syed has told me. He made a small fortune in the oil futures market; it happens all the time—end of the story.

The beer is chilled and inviting. I take a couple of generous swigs and say "This is quite good; just what I need. It hits the spot." I often pepper my language with sayings like this. They gave Syed a chance to query me as to their meaning if he does not know or, with increasing regularity, to explain them to me.

"Ah, yes," he says, "it is just what you need, like the 'sweet spot' in your baseball."

"You are becoming an expert Syed. Soon you will be teaching me!"

I have by now finished most of my beer and am "feeling no pain." I'm not about to use that expression with Syed however. It is sure to invite another exposition on the idiosyncrasies of the English language.

Picking up an earlier thread of our conversation, Syed continues: "You spoke of basic research. Is there any special area you wish to pursue?"

Syed seems genuinely interested in this field of inquiry and I am reminded that, in addition to his formidable background in Egyptology, he has studied mathematics as well.

"I am working on an idea in the area of cosmology. The current thinking about the origin of our Universe is based on a concept called the Big Bang, the prevailing cosmological model explaining the earliest period of our Universe. As the theory goes, the entire Universe expanded from a high-density state to what we observe today. The current laws of physics, however, breakdown at this initial state, referred to as a singularity. Estimates place this event at roughly fourteen billion years ago. What happened at or before the Big Bang is the key to understanding what followed. This theory, while generally explaining the evolution of the Universe from this initial state, fails to satisfactorily account for the initial conditions in the first instance."

Syed hasn't flinched so I take that to mean he is following me, so far. "I am working on a different approach, a different theory. As I said earlier, quantum mechanics and general relativity are not compatible. For example, neither theory explains the presence of what we call 'dark matter,' a mysterious, invisible, and as yet unknown material that comprises the bulk of matter in the Universe. This unknown substance—if that is even the right word to use—is inferred from its gravitational effects on the visible matter that we can see. To my way of thinking, since the Big Bang theory naturally follows from Einstein's general theory of relativity, and the general theory does not explain dark matter, something has to give. So, I am looking at the way dark matter is distributed throughout the universe to see if I can find some way to either modify the general theory or explain dark matter differently—perhaps dark matter itself, like gravity, is more a consequence of unknown forces or space-time curvature than previously believed."

"That is very interesting," says Syed. "And this new theory of yours; it will eliminate the 'singularity'?"

"Possibly. But I still have further details to work out. These ideas, by the way, rely on the work of many others; the theoretical physicist David Bohm, for example. He passed away some time ago but his work has received renewed attention in scientific circles. I am also aware that Saurya Das, a theoretical physicist at the University of Lethbridge in Canada, and a colleague are also working on this problem. Their work might provide new insights that will help us solve this puzzle once and for all; that is if I do not beat them to the punch."

"Now you think you have gotten me on that one," says Syed, with a twinkle in his eye. "You mean that you will solve the problem before they do, correct?"

"Correct."

"Well, I wish you luck, my friend. If anyone can solve the problem, my money is on you." Syed then abruptly changes direction and says, "Are you up for a game of billiards, Jake? Surely you Americans enjoy the sport."

I know I am in my element now. In my early teens, I used to visit a local pool hall with the money I had earned from summer work. I would practice day after day, without my parents' knowledge, of course, showing scant improvement. Then one day an older gentleman—come to think of it, he would have been about my current age—asked if I wanted to play a game of "straight pool." In this game the first one to reach a previously specified number of pocketed balls—calling the sequence of each shot and intended pocket—wins. My skills went through the roof! Not right away; it took most of the summer. Slowly but surely, however, my game improved until one day, I won! We were playing to fifty points and I barely managed to eke out a win, but win I did. If you want to do anything well, I learned, compete above your level. This is going to be a piece of cake!

"Sure," I say. " 'I consider that the hours I spend with a cue in my hand are golden.' " The quote from the Broadway musical, *The Music Man*, which I thought sounded clever. Having spent—wasted? —my youth in pursuit of the art of six-pocket, I am sure I can give Syed a good game; that is until I see that the table doesn't have any pockets. I had taken the word "billiards" in its generic sense. I am beginning to feel out of my league.

"I'm used to playing pool on a table with pockets," I say. "I'm not sure I can manage a good game. I don't know the rules—"

"It is child's play; don't worry, I'll explain the rules. I am sure a scientist can follow them; it is very straight forward." Syed then launches into an explanation of the rules for "three-cushion billiards" or "three-cushion carom." As he is explaining the rules, I am doing my best to internalize the details: The cue ball caroms off both of the object balls and hits the rail cushions at least "three times." Your cue ball doesn't have to contact three different cushions as long as they have been in contact at least "three times in total." A point is scored for each successful "carom," and so on.

It is a disaster! I can't manage to score a single point or "carom." I know the geometry and physics of billiards. What I lack is the "procedural memory" for this specific type of game; skill sets do not, necessarily, transfer so easily. Syed graciously lets me off the hook, suggesting we retire for the evening. "We have a lot to do tomorrow," he says, "You will need your 'beauty sleep.' "

Sounds like a plan; the single beer has taken its toll and I am feeling fatigued once again. Ordinarily, a beer or two would have sharpened my game, had I been playing a "regular" pool game, and had a good night's sleep. I had discovered long ago that there is a fine gradation away from optimum performance after about two beers. Beyond that limit, even the fine-honed skills of an expert decline in rapid order. I will sleep well tonight.

Chapter Five: The Discovery

I wake to the sound of the call to prayer by the local muezzin, the person who recites the adhan-azan from the mosque. It is early, very early. So much for a good night's sleep. This unsolicited wakeup call is the first of several such calls that will be heard throughout the day around Cairo—at dawn, noon, mid-afternoon, sunset, and nightfall. I have no right to complain; after all, Egypt is a Muslim nation and I am the interloper.

Cairo richly deserves its reputation as the "City of a Thousand Minarets." I had noticed many mosques dotting the skyline during our drive in from the Cairo airport. There had been a brouhaha over the cacophony generated by the muezzins issuing the calls to prayer in the thousands of mosques spread throughout Cairo. This tradition changed dramatically when the Religious Endowments Ministry, which oversees the Cairo mosques, initiated the "Unification Project." Under new guidelines, the calls to prayer are consolidated into live broadcasts to be issued simultaneously from a single radio station and spread throughout Cairo by wireless receivers placed in every mosque. Some muezzins, either fearful of losing their position or out of defiance to a perceived assault on a long-standing tradition, simply pull the plug on their receivers during the broadcasts and proclaim the call to prayer the old-fashioned way. After finishing their recital, they plug the receivers back in to remain "compliant"—no harm, no foul.

I get out of bed and head for the bathroom. My trusty travel kit, which includes a 110/220 voltage electric razor, travel plug adapter, sunscreen, and sunglasses, has served me well on several previous trips to Europe. I brush my teeth and then hop in the shower, glad to enjoy the amenities. My stomach reminds me that I have not eaten a good meal since before I boarded my flight in San Francisco. Whatever Syed has in mind for breakfast will be okay by me.

I put on my robe and step out onto the balcony which faces toward the east. My room is upstairs in the northeastern corner of the residence, opposite the courtyard in the back. Syed's room, on the southeastern corner, also has a balcony looking out toward the east as well. It is still dark out and according to my reckoning, the sun won't peek over the horizon for another hour or so, give or take. I decide to sneak downstairs and read for a bit while I wait for Syed to join me.

As I walk down the hall toward the staircase, I notice the downstairs lights are on. There is some noise coming from downstairs but I don't see anyone. I head down the stairs and enter the main room where Syed and I had sat the previous evening. I sit on the sofa by the window overlooking the garden.

Syed emerges into view from around a wall that separates the main living room from the kitchen and dining areas. I am surprised to see him up so early but I needn't have been; he usually gets up early before Morning Prayer. "Good morning, my friend," he says, remarkably bright-eyed and alert for so early in the morning. "I was just about to make some breakfast. I didn't think you would be up so early. Come and join me. We can talk while I am preparing the meal. I am making for you what we call 'Ful.' It is a traditional meal in Egypt usually served for breakfast. I am sure you will like it; it is very tasty. I think you will be pleasantly surprised."

I join Syed in the kitchen and watch in amazement as he begins orchestrating the ingredients for our breakfast. He works like a master chef, constantly moving from one step to the next. "I envy your culinary skills," I say. "I usually settle for a bowl of oatmeal or cereal."

"It relaxes me," says Syed, "and helps clear the mind. I used to watch my mother cook when I was a child and prefer cooking for myself. I could of course afford to have a cook and other servants to assist me, but I prefer my privacy. I have a housekeeper, a Nigerian woman from an agency who comes in once a week to help with the cleaning. Her name is Abasiama. She is from Kaduna. As you know there continues to be much strife in many parts of Africa and Nigeria is no exception. She came here, like so many others, to escape the continuing struggles there. These people continue to quarrel over their chiefdoms, who is in power and whatever else they can find to argue about."

Syed's dismissive attitude toward a large swath of Africa is telling. It is clear he brooks no defense for the violent nature fostered by the divergent regimes in Nigeria and elsewhere; perhaps including Egypt as well. His gentle rant however does more than merely inform me of his political take on the conflicts throughout Africa, it also speaks to his mental framework or approach to problems in general. He holds a strong position on political matters; something for me to keep in mind. Note to self: Steer clear of contentious politics.

"Have a seat Jake; it is almost ready; a little olive oil"— Syed's voice trails off as he continues talking to himself: "Where is the pita? Wait a minute—"

I walk over to the dining room table and take a seat near the end. The table is conventional, about six feet long, with dark rectangular sections on the top bordered by white stripes, with six brown and white straight back chairs tucked in along the edges. I notice how its plain design contrasts with the balance of the décor in Syed's home.

Syed brings our breakfast and places the plates on the table. He then sits down to my left at the end of the table instead of sitting on the opposite side just across from me. I think nothing of it for a moment but then remember that in Islamic countries the host customarily has his guest sit to his right. Syed's gesture is probably due more to habit than any sense of propriety. I pause a moment before partaking. Muslims typically say a silent personal prayer, the du'a, before and after meals, in recognition that all blessings come from Allah.

"Well, don't keep me in suspense; tell me what you think." Syed is eager to get my take on his version of "ful."

I am hungry enough to dive right in but mindful of my manners, so I take a small bite, savoring the flavor. "It's quite delicious," I say, meaning every word. "You missed your vocation, Syed; you would make a great chef!"

"I'm glad you like it. It is a favorite dish in my country. Have some bread." Syed then quickly gets up from the table. "I forgot the tea; I'll be right back."

Syed brings us some hot tea and sets it down. "I added a little milk; I hope not too much. I like to have tea after breakfast. Try it; it goes well with what you Americans call our 'been stew.' Let me know if you need some more bread."

The tea is steaming hot. I take a tiny sip.

As we are eating, Syed casually mentions the tablet. "We will look at it after we eat," he says.

I am stunned! "You have the tablet here; in your house?"

"Where else would I keep it?" he says, taking a bite of pita bread. "My office at the university is not good; it is too open. People come and go as they please. I am not even there most of the time. It is secure; do not worry."

I am impatient to see what the tablet looks like. Syed and I had agreed we should not send any photos of the tablet, not that anyone would necessarily know how important it is.

After we finish our breakfast, Syed gets up from the table. "I'll be right back," he says.

I stay put where I am as Syed didn't say anything about moving into the living room. He returns with the tablet, sets it down on the dining table, and begins busing the dishes. His nonchalant manner is surprising given the import of the find.

Here I am, seeing the impossible, the Schrödinger equation embedded in the tablet. The background is nondescript and fades away as I zero in on the singular equation:

$$\hat{H}|\Psi\rangle = E|\Psi\rangle$$

The formulation presents what has come to be called the general "Time-independent Schrödinger equation." I have seen this representation before on the backs of t-shirts worn by some of my students on campus; their attempt at currying favor in my eyes, no doubt.

I am used to solving Schrödinger's equation to obtain the wavefunction; a mathematical quantity containing all possible information about a quantum system. To do this, one needs to incorporate several specific values, such as Plank's constant; a term called the del-squared operator; a mathematical quantity (i) called an imaginary number; a vector which describes the forces acting on a particle and; of course, the wavefunction itself, denoted by the Greek letter Ψ (psi). Once one adds the appropriate mass for the particle being described and solves the partial differential component describing how the wavefunction changes its shape with time, you're done! You essentially know everything you can know about the quantum system you are examining.

In truth, there is much more to this process. Depending upon the subject under investigation, all kinds of mathematical chimeras emerge—Hamiltonian operators, de Broglie wavelengths, electron spin; along with their related nieces, nephews, and cousins! The ubiquitous "Heisenberg Uncertainty Principle" lurks in the background . . . always. It can be, and often is—to use the technical term—a mess!

Syed returns and sits down again. "Well, what do you think?" he says. "Are we on to something?" If he is as excited as I am about "the discovery," his expression gives nothing away.

"We shall see," I say, distracted by the sheer inexplicability of what I have been presented. I suddenly remember I have left my list upstairs. "I'll be right back. I need to get something."

I hustle upstairs to my room and retrieve the list. As I hurry back downstairs, I see Syed carrying the tablet into the living room. He places the tablet on the table in front of the sofa. "Come on over," he says; "Let us have a better look."

"What are the dimensions?" I ask, curious why the tablet seems so large compared to the inscription.

Syed hesitates for a moment, thinking. "I measured it before . . . let me see. I think—no, that's not right. I'll be right back."

The equation inscribed on the face of the tablet appears to be centered, as far as I can tell. Syed hands me a metric ruler and I attempt the measurement the best I can.

Seeing me fumbling with the meter-long ruler, Syed says: "You Americans are ahead in many ways, but you still refuse to use the international system of measurement."

"I use it every day," I reply, quick to defend my knowledge of the metric. "But it usually involves mathematical computations and not a physical measurement. Give me a moment, I'll get it." I do my best to measure, converting in my head to English Units as I go along; not easy without pen and paper. "Okay, about one foot by one and a half feet. Is this typical for a tablet from the period—what was it again . . . the dynasty?"

"The Early Dynastic Period, from roughly thirty-one hundred to about twenty-six hundred BC. The period is not 'written in stone,' as you say. It varies depending on the scholar who you are listening to. We date the age of the plate closer to the end of the thirty-first century BC and the beginning of the thirtieth. As for your question, whether it is standard, I can tell you there is no such thing. Its size is not unusual in any way, not really—"

"Well, Syed, it looks like we have a job to do." I look at my list and check off a few items. I then ask Syed what I hope is not a stupid question: "How were these inscriptions made, as far as you can tell? I mean, can you determine what type of instrument may have been used? It doesn't look like it has been chiseled; it's too smooth. It looks more like an impression than a chiseled piece of—"

"No," says Syed. "It is not clear how the impression was made. There are no distinctive, typical striations that would indicate the use of tools indigenous to the period but—and this is important—there is no sign that the impressions were added recently. If the tablet had been tampered with, we would find exposure rates to differ from the underlying surface. This would tell us the markings were introduced more recently. Everything is verified Jake; of that I am sure."

Wow! Syed's explication was succinct, erudite, showing an impeccable command of the English language. He does have his moments. I rummage through my mind for a moment searching for answers. "So, the imprint just appears on the tablet and we don't know how it got there," I say, thinking out loud. "We also don't know what mechanism was used to embed the equation into the tablet. I guess we'll just have to leave it at that for the moment." Something else to put on my mental self. The other items on my list will be resolved in due time; many of them now seem less important. Syed is telling me everything he knows and I trust his knowledge and judgment.

"Are you saying there are no markings on the tablet? Is that what you mean?" I am careful not to press the issue too firmly. I don't want to irritate Syed by this line of questioning, but I am confused and need further clarification.

"Yes, that is exactly what I mean," he says. He then pauses for a moment. "Jake, tell me honestly, what are you making of this? It is not making sense to me . . . none."

Taking a scientific approach, I say: "Several questions come to mind, but we have to first make some assumptions, form several possible hypotheses. Some of these assumptions, if true, will lead us in one direction or conclusion, and others may lead us somewhere else entirely." I am doing my best to lay out my thought process intelligibly. I could use Linda's perspective.

I begin, tentatively, to formulate my thoughts: "We have three, possibly four, possibilities—it is a hoax and we are not clever enough to figure out how it was done; it was created during the period suggested and we need to understand why; or, somehow, it represents a message or clue from the future sent back in time to alert or inform us about who knows what. The only other possibility is human time travel but that would be an even more daunting proposition. I haven't ruled it out altogether however and remain open-minded." Maybe Syed has come up with another explanation.

"Your second scenario is very interesting." Syed places the tips of his fingers together for a moment, thinking. "What if the ancient Egyptians, or others, had this knowledge some thousands of years ago. What could that possibly mean for us today? And should not we expect to find additional evidence of other artifacts of a similar nature?"

Syed's comment reminds me there might be other tablets to examine if we can retrieve them. Syed had mentioned this possibility before I left for Cairo.

"Anything about the plate, besides the inscription, that merits attention?" I ask, the thought popping into my head.

"Yes, there is something I should mention. It is unusual. I did not know what to think of it so I had put it out of my mind."

This is a news flash. I wonder why Syed hadn't he told me earlier? He has my attention; I am eager to hear what he has to say.

"When we examine an article like this, we use more than one method of authentication. When I had the tablet scanned it showed gaps throughout the stone. I was quite clever you might know in having this work done so that no one would be suspicious. You would have been proud of me, Jake. I told the technician a phony story about why I needed the work to be done and he ran the test and did not ask me anything. He had no idea what the symbols might mean. These spaces we found are what you would call an 'anomaly.' The technician said he did not think anything was wrong with the scanner and I did not want to attract attention, so—"

"This could mean something," I interrupt, excited but having no clue where a further investigation might lead. "I don't know if it means anything or not just yet, but we should remember to follow up. It could be important."

It is difficult to keep my mind from conjuring up another conspiracy theory. In my attempt to fight off these intruding thoughts, I lose my original train of thought entirely. Like a pinball machine, our minds flit from one thought to the next. Recovering my concentration, I say, "What about this Khaled fellow; are you still in touch with him?"

"He is very secretive about what he is doing. I am sure he has something to hide." Syed speaks in a subdued tone, giving me the impression I am watching a classic detective drama.

"When do you think we will be able to follow up and—"

"We agreed to meet tomorrow evening at the Sangria. It is a club just off Nile Corniche, a short drive from here. If he has something to offer, we will listen. It has a lovely view of the Nile with an outdoor seating area and music. I think you will like it and it will give you a chance to see some of the nightlife in Cairo."

Listening to Syed, I am beginning to get caught up in the pursuit. This could turn out to be very interesting, I wish Linda could be here. She would get a kick out of all the intrigue: A dimly lit nightclub; a clandestine meeting with a nefarious criminal; and the continuing challenge of unraveling a dark mystery! Okay Jake, take it easy. Stay focused.

"I will call Emile," says Syed, "just in case. We may want to follow up after our discussion with Khaled. Besides, I hate driving in traffic."

"So, Emile is not on your staff?"

"He drives for me when I do not feel like driving or I have some function to attend. He has his car and works as a chauffeur to supplement his small pension from his service in the French military, but he is always available if I need him."

"I sometimes forget there is a French military," I say, immediately regretting my flippant remark.

Syed looks at me for a moment, his expression restrained. "The French have many troops in Africa. Mali and other regions have seen French involvement for some time and their presence is well known. Emile Morand has been out of the military for some time but he continues to follow the news reports regarding French military operations. He speaks with great fondness of his time serving the French government. He even volunteered to serve in Operation Serval in Northern Mali during the conflict there some years ago. They did not take him because of his age but he remains fit."

"He seems to be in fairly good shape," I offer.

Syed wrinkles his forehead, looking annoyed. He has no desire to continue discussing French military affairs or Emile's physical endowments. His earlier excoriation of the conflict in Nigeria left me wary of pushing his political hot buttons any further.

"I would love to see the Cairo University where you work," I say, tactfully changing course. "Would that be alright?"

"Yes, of course; I was intending to show you around. Tomorrow morning would be a good day to do that. I have some business to attend to presently. It will only take a couple of hours or so, and then I must see to prayer as it is Friday. I hate to be a poor host and neglect you—"

"No, don't worry Syed, take care of your business; I'll be fine."

"I should be back around one or so, and then we will have some time together. Sound like a plan?"

"Okay," I say, and off he goes.

I doubt Syed would be conducting business on a holy day Friday but I say nothing. Perhaps he is going to see a friend. I'm the guest; it's none of my business.

Early morning Cairo time would be yesterday evening in California. Sometimes today is yesterday, sometimes tomorrow, and sometimes somewhere in between. It makes perfect sense. Linda should be home. I'll give her a call. She's probably doing some last-minute packing for the trip. Should I tell her what she's missing—our clandestine rendezvous with the criminal element?

What is time? Millennia have passed since we first began to tackle this question and we are no more certain as to the real nature of time than the ancient philosophers Heraclitus, Aristotle or Plato were. Did Einstein help or hinder us? Confuse is more like it. Sometimes, rather than arguing with myself, I imagine a conversation between two of my imaginary friends, Thomas and Richard (I'm a scientist; I'm entitled—I'm exempt from growing up!). Once in awhile—but only once in a while—Harold has something to say and, out of courtesy, I listen:

Thomas: "Suppose I have a fish tank and it is full of guppies and one is expecting."
Richard: "What color is it?"
Thomas: "Bluish, with flecks of gold in its tail. What difference does it make?"
Richard: "Okay, go on."
Thomas: "Well, if the mommy guppy has her baby, does the water level rise?"
Richard: "No, I don't . . . think so . . ."
Thomas: "Explain."
Richard: "There is only so much water in the tank, so surely the amount of water stays the same."
Thomas: "Sure, but haven't you added a guppy? What happens then?"
Richard: "Wait a minute. Wait just a minute. You haven't *added* anything; you've only redistributed the total 'guppy-mass' in the tank. Nothing was introduced from the outside."
Thomas: "Exactly. There is a finite amount of water. We can neither add nor subtract any water from the system if we maintain that the fish tank contains all of the possible water there is or can ever be. So how much time is there? Can we add to or subtract from the total amount of time there can ever possibly be?"

Richard: "I think you're trying to trick me, but I don't know how yet."
Thomas: "Now, would I do that? Never mind, the point is: If time is a definite thing—like Jell-O or—
Richard: "Strawberries?"
Thomas: "Okay, strawberries. Quit interrupting . . . Where was I? Oh yeah; so, if we can't make time—we can't go and get some more—there is only so much of it, then, piggybacking off of the first law of thermodynamics, whatever amount of time there is, that's all there ever will be."
Richard: "Just like the water in the fish tank!"
Thomas: "Exactly."
Harold: "But where does time come from in the first place?"

And so it goes. When Sir Michael Philip Jagger sang the lyrics— "Time is on my side, yes it is"—half a century ago, it was . . . and now . . . it isn't. What happened? Did he age *because* of time? Did time *do* something to him? Or did he simply pass *through* time?

How old is anyone anyway? Our cells are completely replaced every seven years or so, give or take (Or so we're told. I should ask Linda; she'd know). So, when I first meet someone, depending on what point in our cell-replacement-cycle each of us is in, we could be the same age or somewhat older or younger than each other. How crazy is that?

Waxing philosophical, as I am wont to do, I could argue that everyone and everything is the same age—fourteen billion years old, the approximate age of the universe. Our current physiological existence may have been introduced more recently, but the ingredients of which we are composed either came into being at the same time or have always been here.

When we push the "pause" button on our television or streamed programing, do the characters on the screen notice? Do they wonder who or what froze them in time momentarily? Of course not; even if the program is live when we interrupt the feed and then restart the program, it continues as if nothing happened. We're now behind the characters in terms of the action, but by the time the program ends, we have all of the information about the episode; same as anyone else viewing the same program. If our universe were put on "pause," would we notice? Our thoughts would be frozen as well . . . wouldn't they?

Dr. Oliver Sacks, in his classic account of the encephalitic lethargica epidemic that occurred just after World War One, depicts patients frozen in time. They age, decade after decade, but show no signs of perception; no indications whatsoever that there is a conscious, thinking person inside. After Dr. Sacks administers a medication called L-DOPA in a carefully measured dosage, the patients, after experiencing decades in a "living coma," spring to life! Their minds were locked inside their skulls—awake, alert, but unable to communicate with the outside world. What must time have been like for them?

The proverbial tale of twins—one remaining on Earth while the other, an astronaut, speeds off on a long voyage traveling near the speed of light—is familiar. If the astronaut twin travels at eighty percent the speed of light for ten earth-years, upon his return he will find that he has aged a mere six years; whereas, his stay-at-home twin will have celebrated ten birthdays. The discrepancy in their respective clocks is due to time dilation, a difference in elapsed time as measured by observers moving relative to each other. The calculation is quite simple; a junior high school student could complete it with a calculator in no time. It's the derivation of the equation, not its simplicity that matters. Did each twin experience time differently than the other or did one of the twins undergo greater change than the other? Is there a difference?

Psychological time is quite different than clock time. I remember reading about a man who had been blind from birth. He plays golf, with a little help from his friends. He wrote of a time when he was sitting by the backyard pool with his two-year-old son. His wife was away on a short errand and expected back in a few minutes. He was sitting on a lawn chair relaxing when he heard a "plopping" noise. His heart sank! His son had fallen into the pool. He immediately jumped in the pool, froze every muscle in his body, remaining completely motionless, and listened. Sensing a tiny disturbance in the water as his child moved through the fluid and hearing a muffled sound, he swam directly toward his son and pulled him out of the pool. Fortunately, his son was more confused than harmed and quickly recovered. When later recalling the traumatic episode, he mentioned how his mind had distorted his perception of time. To him, the incident's duration seemed more like an hour than the few moments that passed. Was he experiencing psychological time dilation?

Some scientists view time as having originated at the Big Bang; others that time has always existed; and still others that time is merely an illusion or mental construct—Einstein embracing the latter view. Specifically, Einstein believed that the "separation between past, present, and future is only an illusion, although a convincing one."

To believe that time is a material substance like Jell-O or strawberries is of course absurd, but the comparison serves to point out the essential problem we face when attempting to explain exactly what time is. If it isn't a substance then it must be something else—but what, exactly? Julian Barbour is a British physicist who also believes that time does not exist. He takes his argument further by asserting that many problems in physical theory result from our naïve assumption that time does exist. Our memory is the only evidence we have of the past, he tells us, and evidence of the future relies solely on belief. According to Barbour, it is change that creates the illusion of time. Each moment stands alone, complete—it does not come into being.

My problem is to figure out a mechanism that could reach backward in time to alter the material integrity of an object. If time is an illusion, perhaps it isn't necessary to reach across space and time; maybe all time—or as Barbour would say, all "nows"—is available to us anytime we choose. Could a future civilization have access to what we call the past, present, and future? If time is like a deck of cards, each card representing separate "nows," an intelligence (perhaps even ourselves in the future) might be able to simply "grasp" and alter one of these "cards," leaving it for Syed's subsequent discovery.

I decide—or my subconscious or whoever is in charge of the cortical space between my ears decides—to just call Linda and let my thoughts unravel as they may.

"Hey," says Linda, sounding relaxed and cheerful. "Everything's all set; my flight leaves at two o'clock tomorrow afternoon."

"That's great. The weather is in the 'Goldilocks zone.' It should be in the mid-seventies today; just like back home." She will learn about the air quality soon enough without my telling her. "Syed showed me the tablet—"

"What's it look like?" asks Linda, cutting to the chase, as usual.

"Well, that's the thing," I say. "The actual impression on the tablet, as best I can make out and from what Syed tells me, wasn't carved in the way you might expect—"

"What do you mean? Is there something unusual about it?"

"Syed tells me it hasn't been tampered with and I believe him. He says the inscription—and this is the weird part—just appears in the stone. It doesn't show any signs of having been carved into the tablet at all. The best way I can describe it is . . . well, the equation—the impression it makes—is simply displacing the material on and just under the surface; if that makes any sense."

"If it were stamped into the tablet there would be signs of compression, some obvious surface alteration—due to pressure or heat of some kind."

Once again, despite my imperfect delivery, Linda has grasped exactly what I was trying to say. "Something else to figure out, I suppose. Oh, I almost forget—Syed tells me there are 'gaps' in the tablet; at least that's what I call them. The anomaly was discovered when he had the tablet scanned."

"Wasn't the technician suspicious?" asks Linda, concerned about a possible breach of security.

"Syed told me he was able to pull off a ruse and the technician didn't suspect anything. I didn't press him about it; he seemed confident enough that it wouldn't be a problem."

Linda's interest in the tablet is genuine; she is as eager to get to the bottom of this mystery as I am. She also wants to play a role in the unfolding drama of discovery; and yet, there she is—a continent away—left out of the hunt, getting a verbal report by phone. Her originally cheerful mood gives way to one of frustration. "What about the other tablets?" she says, probing further.

"We're going to meet with Khaled tomorrow evening—"

"What time?" asks Linda, sounding increasingly anxious.

"Syed didn't say; tomorrow evening sometime." Wait a minute! Linda will be arriving tomorrow! Get your head together Jake! "What time is your flight expected?"

"Well, it leaves at two and there's a stop in Frankfurt—"

"So, it's pretty much the same as mine was. Let's see, two o'clock, plus twenty and a half hours . . . hold on a minute . . . midnight would be ten hours, plus another ten and a half. That's ten-thirty in the morning Pacific Coast time; add nine hours . . . You should be here around seven-thirty or so if everything goes perfectly. That's about right with the layover and connection from Frankfurt. I figure the whole trip is about twenty and a half hours, give or take."

"How did you figure that out so quickly? Never mind; if I get there before eight tomorrow evening, I should be able to go with you to meet this guy, Khaled."

"Trust me, the last thing you'll want to do after twenty hours of travel is to go to a night club."

"You're meeting at a night club? I didn't think Syed drank; does he? Anyway, I'll be alright. I'll just have to try to get some sleep the best I can along the way."

I know Linda is just trying to be positive, or maybe she is just rationalizing. There is little chance of her getting much rest during the trip. Either way, I know she will insist on coming with us. Besides, we will need to pick her up at the airport. It wouldn't make sense to take her to Syed's home and then rush off without her to meet Khaled. She wouldn't go for it—no way. I hope she can make it in time so we could share the experience.

"Did you get a chance to look up any information about Cairo?"

"I know, dress modestly, cover my arms up to the elbows, dresses below the knees. I found a couple of nice Maxi dresses at Macy's. They have a diagonal blouse overlay. I thought I looked pretty good. Wait a minute; are the sleeves long enough?"

"You would look mah-velous in anything," I say, affecting a French accent in a feeble attempt to sound debonair.

Ignoring my attempt at flirtation, she says "I brought a couple of pairs of pantsuits but I'm not sure what to wear. I'll figure it out. I also bought a couple of scarfs, just in case. And, I know, offering to shake Syed's—excuse me, Dr. Azad's—hand may be a little tricky but I'll just play it by ear. Oh, and I thought this was interesting—don't ask for salt on your food; it's considered an insult to the chef. I think I'm pretty much up to speed."

"Hmmm . . . I didn't know that about the seasoning. I don't remember using any salt with my breakfast. I did know enough to take my shoes off in the house." Switching gears, I ask Linda what she makes of the 'gaps' that Syed's technician discovered in the tablet.

"I don't know; that's pretty strange. I'll look into it if I get a chance before I leave. There's always better technology we might use to see what's going on."

I trust Linda will figure it out or find out what to do if she can't. She's better at research than I am. "I'm glad we're doing this together. Give me a call when you reach Frankfurt."

"This is starting to get exciting Jake. I can't wait to see you. I'll call you soon."

So, she can't wait to see me, huh? Did she mean it or is it her anticipation of an adventure that made her say it? No matter, whatever excitement we encounter, we'll be together. It might bring us closer and that would be okay by me.

Chapter Six: Strange Bedfellows

Linda and I work out several scenarios that might explain how an ancient Egyptian tablet bore a twentieth-century equation. Neither "Occam's razor" nor Sherlock Holmes's famous elimination of possibilities is of any use to us under the present circumstances. The tablet's very existence introduces rather than eliminates the "impossible." Razors are of no help either as our quandary has no "simple" explanations from which to choose.

Deductive and inductive logic—moving from the general to the specific and vice versa—are of no help. Having but a single tablet to work with, any attempt to generalize would be woefully inadequate. Likewise, the same problem hinders any attempt to go in the opposite direction—from the general to the specific.

How do we solve problems? We use the tools available to us—our minds, technology, whatever clues we have. We use these tools, combining them in different ways, to produce a solution to a problem or solve a puzzle. Sometimes what's missing becomes the clue we need. A dog that didn't bark provided just the clue that Sherlock Holmes needed to solve a case. Could the gaps, hidden in the tablet, be the missing clue we need to solve our mystery?

Syed returns invigorated; his visit to the mosque renewing his energy. "Let me make us some lunch. I have some falafel and pita bread that I can prepare in no time at all. Are you thirsty? You found the bottled water—"

"Yes, I'm fine." Syed is an excellent host. He seems genuinely interested in making me feel comfortable in his home. I know him professionally but as we have come to know each other on a more personal level, I feel a growing bond of friendship.

I mention to Syed that Linda should be arriving tomorrow evening and is hoping to go with us tomorrow when we meet Khaled. I figure I might as well be direct with him. He hadn't voiced any complaints about her involvement when I first mentioned her coming aboard.

"What did you say is her subject?" he asks.

"She teaches molecular biology. She has done a great deal of work in genetic research. She has a very analytical mind and will be an asset to us." I am hoping to boost Syed's opinion of her any way I can.

"What time is our meeting with Khaled?" I ask, quickly moving the conversation along and away from Linda.

"We agreed to meet around eight-thirty or nine. When is Linda? What is her last name?"

"Cooper."

"When will she be arriving?"

"About seven-thirty or so, if there are no snags."

"Well, if there are no 'snags' as you say. Let us have something to eat."

I'm not about to add anything to Syed's matter-of-fact replay. "Sounds good."

After lunch, Syed and I sit outside in the courtyard enjoying a cup of tea.

"Tell me about your recent work, Syed. Any new digs lately?"

"I had an opportunity to work with a team examining the tomb of the Egyptian god of death, Osiris. It was discovered in what we call the Valley of Nobles in Luxor."

"Where is that?"

"It is about five hundred kilometers or so south of Cairo. The find is believed to date from the twenty-fifth or twenty-sixth dynasty; perhaps as old as the eighth to the late sixth century BC. Luxor is an interesting area for such finds. The French-Egyptian Centre and others have been active in this area for some time."

"It must be fulfilling to make such contributions in the field. I used to think I would make some great discoveries or contributions to my profession. I've accomplished some good, I suppose, but I have never had the satisfaction of making a breakthrough in science as so many others have." After telling Syed about my failings, I wonder what he must be thinking of me.

"Have you seen the American movie, *The Bridge on the River Kwai*?" he asks.

"Yes, it's one of my favorites."

"It has two characters in the movie that are very good, William Holden and Alec Guinness. Mr. Guinness plays the part of Lieutenant Colonel Nicholson. There is a scene where he is speaking to the Japanese commandant and reflecting on the meaning of his life. I have memorized his words as they remind me I feel the same way sometimes:

> There are times when suddenly you realize you're nearer the end than the beginning. And you wonder, you ask yourself, what the sum total of your life represents. What difference your being there at any time made to anything. Hardly made any difference at all really, particularly in comparison with other men's careers.

"I sometimes find myself wondering the same thing, Jake. Whether my life will have any meaning. Will it make any difference—as Mr. Guinness said— 'really'? I want to believe that my life will matter; that I will have 'made a difference.' "

This is a side of Syed I have never seen before. I am deeply moved he would share such personal feelings with me. "Syed, if you will permit me, I have a little story that you might find interesting. It is a moral tale of Sufi wisdom. I think you will find it instructive."

"Please, I would love to hear it."

"Well, there is a man and his grown-up son who are living just outside their village. The son comes home one day and with great excitement tells his father he has found a horse. The son tells his father there are no markings of ownership on the horse and it will be of great benefit to them. 'Is this not good news father?' he says. His father only replies, 'We will see.' And then, riding the horse one day, the son falls off and injures his leg and is unable to help his father with the chores. 'It is terrible' the son says; 'I am not able to help you. Is it not very bad father?' His father, once again, replies, 'We will see.' His son does not understand and believes his father is not thinking clearly. Finally, the military comes to take the father's son into service but, because of his injury, they leave him with his father and go away. 'Is this not a good thing?' his son asks his father. 'Now I can get better and I will be able to help you soon.' His father then replied—"

"We will see," says Syed, finishing the line for me. "That is a very good story, Jake. I think I understand its meaning: Who can say what the outcome of our lives will be?"

"And who knows," I say, "perhaps we will solve our mystery and become famous!" It seems like a good place in our very serious discussion to inject a little humor.

We both laugh heartily. Syed, getting into the spirit, says "Perhaps I will become president of Egypt!"

We laugh again. Syed then comes up with the perfect interlude to our philosophical musings. "How would you like to watch some old classic movies? I have one I have been meaning to see again. It is called, *The Treasure of the Sierra Madre* and is staring Humphry Bogart."

"I would like that very much," I say. We spend the rest of the evening watching old American film classics. I ask Syed if he would like to watch *The Bridge on the River Kwai* again. He is delighted at the suggestion and, when we finally retire for the evening, we are better friends than when the day began.

Chapter Seven: Prelude

I manage to sleep straight through the night, my internal clock calibrating to the time difference. Syed is up—no surprise—and ready to get underway. From my calculations, Linda would be arriving in Frankfurt about ten a. m. this morning Cairo time. That would make it about one in the morning Pacific Coast time so she will probably be a bit groggy from the trip. The best I can hope for is a call from her sometime during her layover stop in Germany.

Syed has made arrangements for Emile to drive us to Cairo University and they are sitting in the living room when I come downstairs.

Emile is adept at negotiating the traffic in Cairo, which is saying a lot. There are few traffic signals and as far as Emile is concerned, they are advisory—he routinely ignores them. I had visited Manhattan a few times and, after my first trip, never again rented a car. I didn't think another city could match Manhattan in terms of traffic. Wrong.

Arriving at the university the first thing that hits me is the enormity of the campus. There is no way we will ever be able to see everything in a single tour.

Syed has Emile drop us off near the dome of the university and we begin a casual stroll through the grounds. "Wow!" I say, "What an enormous campus. How do you ever manage?"

"I am thankful not to be working in the administration. It is a great responsibility. There are many disciplines we teach and it is difficult to keep everything organized and on schedule. And there have been security issues as well. Student political groups have caused trouble for us but we have addressed this situation. We are continuing to make much progress and I am encouraged we will continue to meet the needs of so many of our students."

Syed's marginally candid comments sounded more like a political speech than the casual musings of an academic. We continue our tour, alternating between walking and driving. There are so many students and faculty; I have no idea how anyone could coordinate such a massive complex. I notice women professors on campus which I was told is in no way unusual. Syed shows me the Faculty of Archaeology where he works, explaining that it encompasses ancient history, Egyptology, and anthropology. The Grand Celebration Hall is quite impressive. We manage to sneak in while it is not in use. Syed shows me the translation devices available in the auditorium. He takes a great deal of pride in the university and the role he plays in promoting its success. He mentions a few of the former alumni, among them Omar Sharif. I think immediately of the movie, Doctor Zhivago.

"Is there student housing available?" I ask, making polite conversation. How can a campus of this size not have student residence facilities?

"Yes, it is very much like a little city. We have students pursuing their degrees from many lands, not just in Cairo, so we have to make room for them as well."

Even though we drive around for part of the tour, we do our share of walking and Syed is visibly tired from our excursion. "Come, let us go into the city and have some coffee," he says. He probably just wants an excuse to terminate the tour, unless the coffee available on campus is especially bad.

Emile drives us to the Safir Hotel in El Messaha Square where The Palm's Coffee Shop is located. One of Syed's daytime hangouts no doubt.

Syed insists I try the Viennese roast, referring to a particular variety of beans he prefers when drinking Turkish coffee. Syed orders for us and the waiter eventually returns with our coffees, a glass of water for Syed, and a bottle of mineral water for me. I start to take a drink of coffee when Syed interrupts—

"No, no," he says, "You will want to let it settle for a moment. And have some water before you drink your coffee."

After the coffee grinds have time to settle, I start to take a drink when Syed interrupts again—

"Just take a sip or two and enjoy. You Americans are always in such a hurry."

There are some customs you just have to learn as you go along, I suppose. Syed makes an excellent tutor.

Syed and I are "kicking back," enjoying our coffee and each other's company when I hear the familiar "Jack Bauer" ring.

"Hey, Jake," says Linda. "How is everything going? Don't tell me I missed out on anything."

Linda seems in good spirits, considering. The jet lag will hit her soon enough. "No, you haven't missed anything. I'm at a café having some Turkish coffee with Syed. How is your flight . . . so far?"

"I landed in Frankfurt a few hours ago. I thought I would wait a bit and make sure the connecting flight to Cairo is on time before I called. It's almost noon here right now."

"Sounds about right; it's almost one o'clock Cairo time. So, when should we expect you; around eight or so?"

"Well, let's see; there are two-and-a-half hours before my flight leaves . . . and about four more hours to go—"

"That's about eight or so; maybe a bit earlier. We're an hour ahead of Frankfurt. You'll have to go through the usual rigmarole when you land but, not to worry; we'll be here. Get any rest on your trip . . . 'across the pond'?" I ask, in my usual bungling way when speaking to Linda. Every time I talk to her, I take on another personality—my usual awkward attempt to sound with it.

"I got some rest. The seat next to me was empty—that was nice. I'll be okay. See you in a bit."

"Okay, see you soon."

Syed hadn't been paying attention while I was speaking with Linda. He is just sitting back and enjoying his coffee like he doesn't have a care in the world. I could get used to this.

I have only seen the view of Cairo that has been presented to me. I am well aware that there are many slums around the city where hundreds of thousands of the indigent poor eke out a living; where little children rummage through the garbage; and Sudanese refugees are marginalized and subject to religious discrimination.

Leo Tolstoy tells us in his novel *Anna Karenina* that, "All happy families are alike; each unhappy family is unhappy in its own way." When it comes to poverty, however, it would be unconscionable to say that "each poor family is poor in its own way." The poor in Hildago County Texas may face different challenges than those in Cairo but—once a certain, minimum, subsistence level has been breached—quibbling over such differences becomes pointless.

Scripture informs us we will *always* have the poor among us. Cairo is no exception. But it is far from unique—many slums around the world reflect similar conditions. Pulitzer Prize-winning author, Katherine Boo, documents the human suffering in Annawadi, an impoverished settlement near the airport in Mumbai, India. Haiti and Bangladesh are other countries that come easily to mind. Even the so-called *First World* is not exempt: Hidalgo County in Texas is home to the *colonias*, farm-working families that subsist on next to nothing. England and Australia have their share of the poor. Worldwide, a billion people live in slums. In some countries poverty appears intractable; whereas in others, there is growing optimism that ongoing relief efforts will eventually pull many out of the clutches of poverty. One can only hope.

We head back to Syed's house and have a light meal of chicken and, for dessert, "konafa with cream." I can taste some cheese and pastry; perhaps a little syrup, in the dessert. He certainly didn't have time to whip this up on short notice; tucked away in the refrigerator no doubt. Syed has thought of everything and is doing an excellent job of spoiling me. Being the guest, I will just have to tough it out—a little extra time in the "gym" when I get home and I will be as good as new.

As we relax sitting in the living room enjoying some tea Syed's phone rings. Instead of taking the call where we are, he excuses himself and walks off into the adjacent game room. I can barely hear his voice from where I am sitting but it is clear he is speaking in Arabic. After a few minutes, he returns and sits down in the "uncomfortable" chair next to me. That's odd. He loathes that "designer" chair. He looks concerned as if he had just received some disturbing news.

"Is everything alright?" I ask, instantly realizing I am meddling . . . again.

"Yes, yes . . . nothing to worry about."

Why couldn't Syed tell me what the call was about? Never mind; it's none of my business. "What do you say we have some more of that wonderful konafa?" I ask, trying to divert Syed's attention to calmer waters. After all, what's "comfort food" for anyway?

Syed, realizing he has allowed his frame of mind to take a wrong turn (ever the host), takes the bait. "Splendid," he says, and off he goes to retrieve more calories.

As we enjoy seconds Syed asks if I am interested in playing a board game called "Senet."

"I've never heard of it. Another game where you will have the advantage I suspect."

"No, not at all— with your towering intellect you will master it in no time at all!" he says, setting me up for a fall, or so he thinks.

"I will explain the rules for you; it is very simple."

Famous last words!

Syed invites me to join him in the game room and gestures for me to have a seat at an octagonal game table just to the right of the entrance. A large cabinet stands next to the table. He pulls out a rectangular box from the cabinet and brings it over, setting it down in the middle of the table. He then sits down and carefully explains the rules for playing Senet:

The top of the rectangular box has thirty numbered squares arranged in three rows of ten squares each. They are numbered from one to ten along the top row, left to right, and then continue with the number eleven under the number ten, reversing direction, ending up with the thirtieth square on the bottom right corner. Each player receives seven 'pawns'—black and white chips—arranged in a specific pattern at the beginning of the game. On each player's move, five coins are tossed simultaneously to determine how far each player moves his pawns—heads counting one point and tails zero. Starting from the fifteenth square, the pawns move along the board until reaching square twenty-nine, the "House of the Re-Atoum." Once on this square, a pawn may only be removed when a two is thrown. The winner is the first player to remove all of his pawns from the board.

Most of what Syed was explaining flew by me, especially that bit about the "House of the Re-Atoum," but I think I got the gist of the rules of play. Since much of the game of Senet involves luck, the flip of coins to obtain the number of squares or "houses" each player moves, I think I have a fighting chance at redemption from my earlier disappointment at billiards.

My novice approach—confused and befuddled—pays off. I win handily. Syed, always the gentleman congratulates me on my victory. I can tell, however, that he isn't altogether pleased with the outcome. Perhaps I should let him win a game of billiards to cheer him up. That shouldn't be hard to do. Nah, never mind; we're even—one game apiece—let's leave it at that.

Syed brings us some tea and we relax in the living room. I notice Syed has the habit of placing a sugar cube between his teeth when having tea. I chalk it up to yet another cultural habit of the Cairenes.

I am hoping to ferret out a few more clues before Linda arrives. Perhaps Syed knows more about Khaled than he has already told me. "This Khaled fellow; do you know very much about him?"

"He is not what you would call a 'respectable gentleman.'" says Syed. "He makes his living by criminal means. When he first contacted me at the university, I asked him how he got my phone number. He told me some phony story. I then asked him why he was calling and he told me he had an 'artifact'—that was the word he used—that I might wish to see. Even though he did not know what he had in his possession, he knew enough to guess it might be of interest to me. So, I decided to see what he was promoting. He brought the plate into my office and when I first saw it, I thought, as would anyone, that it was a fake, a forgery. But as I examined it more closely, I decided that it was worth a closer examination. He offered to sell it to me for three thousand pounds. I told him I would give him twelve hundred Egyptian and he accepted immediately."

"Twelve hundred Egyptian . . ."

"It is not so very much," says Syed. "In your American dollars, it would be one hundred fifty or so . . . perhaps. I'm not sure. I think it was worth it. Do you not?"

"It could be worth quite a bit more if we can figure out exactly what it is and how it got here."

Syed looks at me for a moment, lost in thought. "Kahled Mosi told me there are other plates like this one. I suppose he meant that there are other plates that he also does not understand, like the one I purchased from him. He offered to sell them for more money if I should want to buy them. He would not say how many tablets he has but I should think there are not a great many. Otherwise, he would have tried to sell them to me already. I think also that he does not have them in his possession. He may be working with others and needs to talk to them. It is all very . . . murky."

"What if he wants us to go with him somewhere else? Could that be dangerous?" I ask.

"Yes."

I was expecting a bit more from Syed than "Yes." If Linda arrives in time to go with us to meet this guy, I don't want her to be in jeopardy—or me either for that matter. "Shouldn't we just let him bring whatever he might have to us? Wouldn't that be safer?" I offer, hoping for more assurance.

"We could do that. We would be risking the possibility of the tablets being sold to someone else. Khaled Mosi already knows I have some interest in what he is peddling. He might want to use the tablets as a 'bargaining chip' and threaten to sell them to somebody else if we do not keep a watchful eye on what is going on."

I couldn't argue with Syed's logic; he has a point. Still, it might be dangerous. And the last thing we need is a "price war." We will just have to take the risk. "Could we bring a bodyguard with us?" I say, only half-joking.

"We will be alright," he says. "Emile will go with us. His military training is very good. We will be okay; you need not worry."

That's easy for you to say. Me, I think I'll worry. If there was ever a right time for a "power nap," this is it. I excuse myself, telling Syed I want to rest for a bit, using my trip as an excuse. I go upstairs and lie down. Even though the anticipation of tonight's encounter with Khaled and the prospect of seeing Linda again linger for a while, I eventually manage to doze off.

Chapter Eight: Portal to the Unknown

On our way to the Cairo Airport to meet Linda, Syed is uncharacteristically silent. My thoughts wander: How did all of this get started? Here I am halfway around the world in a foreign land I know very little about, chasing the answer to an unfathomable mystery; dragging Linda, an unsuspecting accomplice, along with me. I'm either on the brink of a monumental discovery or setting the stage for my eventual demise in the academic and scientific communities. Is this what it's like to be bipolar?

Syed had called me "out of the blue" and very casually mentioned that he had a tablet—ostensibly from ancient Egypt—that contained an equation, Schrödinger's equation. Why? Not so much why this particular equation but why Syed and not someone else? Why me and not someone else? Why Linda? From very early on I have had the premonition that, somehow, events are being orchestrated; that my involvement is encoded in a much larger scheme—just as the tablet is evidence of predetermination or foreknowledge. Perhaps I am just a puppet, but if so, exactly who or what is pulling my strings?

We arrive at the Cairo Airport early enough to ensure we will be on hand when Linda's flight lands. It is after eight o'clock in the evening by the time she clears customs. Syed is gracious as usual and extends his hand to Linda in welcome. She responds in kind and addresses him as Dr. Azad. Cultural norms aside, Syed's respect for Linda's professional credentials is evident; as it would be with any number of female colleagues, he works with at Cairo University. If Syed *does* harbor any reservations about Linda's involvement in our investigation, she is determined to help him discard them post-haste.

Linda is wearing a light beige Maxi dress and a matching, long sleeve blazer, the dress sporting a high neckline, trailing sufficiently below the knees to pass muster. Her shoes are black with a solid, low heel that could best be described as "comfortable." She has done her homework.

Syed shows little concern about our impending meeting and gives Emile instructions to take us to his home. What sort of host would he be if he did not at least allow his new guest a few minutes to freshen up? I remember Syed telling me the meeting with Khaled had been scheduled between eight-thirty and nine p. m. this evening. In Egypt, some customs are inviolable; whereas others—not so much.

After a smattering of polite conversation with Syed about the weather and local interests in Cairo, Linda focuses on tonight's meeting with Khaled Mosi. "Do you think Khaled has any additional information that might help us in our search?" she asks. "Or do you think he might just be trying to lure us into some kind of trap? I mean, if he had other tablets, wouldn't he have brought them to your attention earlier?"

Syed shifts uncomfortably in his seat. "Khaled Mosi is certainly not anyone to be trusted," he says. "Of that, you may be sure. I do think however that he does have something more to offer. He has spent much of his life making a living by selling what others are willing to pay for. He does not come into possession of such things by honest means. And we may also be sure he is working with others. We will have to be cautious in our dealings with him. There are a good many people in his position who have, as you say, 'nothing to lose.' We will see what happens."

Linda seems satisfied with Syed's observations about Khaled. She feels the same way as he does about tonight's meeting—caution should be our byword.

Syed has prepared the downstairs bedroom, just below mine, for Linda. Once we arrive home Emile takes Linda's luggage to her room. Linda seems relieved to have a place to rest and recover from her flight. She knows it would have to wait however as we have one last task to complete before she can retire for the evening.

Syed, Emile, and I gather in the living room to wait for Linda to get settled in. Syed is telling me about a trip he had been planning for an upcoming dig when Linda comes into the room. She is anxious to get going and hadn't even bothered to change.

"Well gentlemen, shall we get underway?" Linda, as always, is not one to dilly-dally; she knows what is at stake and is eager to get on with it.

I look over at Emile wondering what he must be thinking. Syed surely hadn't told him anything significant about our enterprise. If he is curious—and how could he not be?—he isn't letting on.

Syed, though appreciating Linda's enthusiasm, makes no effort to give her the upper hand. He simply sits for a moment and then, gradually rising from his seat, nods toward Emile.

It is a short drive to the El Shagara Sangria club. Emile drops us off and we head toward the club entrance. We catch sight of Khaled loitering just outside. He had made a feeble attempt to make himself presentable. He is wearing a dark jacket over a long-sleeved, white shirt with dark trousers that drape just over the top of his shoes. I guess his age to be somewhere in his late forties. His face is heavily wrinkled and his dark complexion, receding hairline, and generally unkempt appearance made it all but impossible to discern his true age. He could be sixty years old for all I can tell.

"I am surprised they have not told him to leave," says Syed, with an obvious air of contempt. "Let us get this matter concluded as quickly as possible."

Linda and I look on as Syed ignores Khaled's greeting. We then go inside the club and take a table outside on the terrace overlooking the Nile River. Syed orders a "Saudi Champagne," a non-alcoholic mixture of sparkling water and apple juice. We each, in turn, follow suit.

Syed had previously spoken to Khaled in Egyptian Arabic but in deference to Linda and me, attempts a conversation in English. Khaled's English is atrocious and Syed is forced to abandon the effort. Speaking in Arabic, Syed questions Khaled and, after each go-around, apprises us of his progress in extracting further details that might help our investigation.

Syed tells us Khaled has one tablet in his possession and knows of the whereabouts of another. Linda asks Syed if Khaled can describe the tablet. All Syed can get from him about the tablet is that it contains some very strange markings; nothing resembling hieroglyphics or any other symbols Syed would associate with ancient artifacts. The best translation Syed can come up with to describe Khaled's impression of the stone markings is "scribbling." It is all gibberish to Khaled, which suits us just fine. He has no idea what he has in his possession, only that it may be of interest to us. He is surely curious however since now three professional academics are showing interest in his find. The way I see it we had better act quickly before he realizes the full import of his ill-begotten booty and demands a king's ransom for it.

Eventually, the matter of payment comes round and, despite our not having seen the tablet yet, Khaled offers to—loose translation—"let it go" for five thousand Egyptian Pounds. The hefty increase in Khaled's asking price from his last transaction with Syed reflects changing market conditions, and he is fully intent on taking advantage of them, and us.

Syed, as before in his dealings with Khaled, counters with a lower offer—three thousand EGP. Unfortunately, Syed's offer gave more away than he thought. By making such a high offer—even if considerably less than the sum requested by Khaled—he has tipped his hand and signaled his increased estimation as to the tablet's worth. Khaled doesn't budge.

I had read that Egyptians are tough negotiators and expected Syed to continue wrangling with Khaled over the price of the tablet. Syed was taken aback by Khaled's obstinacy. I had seen him upset before but this was different. It was as if Khaled had countermanded an edict. Egyptians are known to linger in their gaze when speaking to each other but Syed's laser-like stare adds a new dimension to this cultural habit. He seems genuinely angered and I am afraid he might botch the deal altogether. He didn't. Contrary to my expectations Syed manages to compose himself, agree to the original asking price, and enquires of Khaled where we might collect the tablet. Syed's ability to remain in the present and focus on the task at hand is uncanny; another facet of his personality that I have come to admire.

We agree to meet Khaled tomorrow morning at *The Citadel*, a tourist landmark that Syed believes will provide a measure of safety. Khaled is to bring the tablet and we will make the payment, as agreed.

On the drive back to Syed's home I feel relieved and on edge at the same time—relieved that Linda and I had met Khaled and taken the first important step in solving our mystery and on edge at not knowing what we might next encounter.

When we arrive at Syed's home, he shows us into the living room and brings us two opened chilled lagers, and two glasses. No need to engage in false modesty with Syed; he knows I enjoy a drink and presumes as much for Linda. He then excuses himself and goes outside to speak with Emile for a moment.

I know Linda would have preferred a Cabernet Sauvignon instead of a "beer" but I also know she can kick a few back when in the mood. Syed's entire stock consists of nothing but lager—take it or leave it. We take it.

Linda and I are alone, if only for a moment. I ask her what she's thinking so far about our little caper. Her answer is classic Linda.

"Khaled seems like the sort of guy that could kill you and then sit down for a nice meal," she says, taking a couple of generous gulps of lager. "I hope this Citadel place is conspicuous enough that he won't think of trying anything." This is vintage Linda—upbeat, humorous, with just a touch of paranoia; a woman after my own heart.

Linda and I continue to empty our glasses as we wait for our host to return. "So, what have you been doing these last few days?" Linda asks.

"Syed showed me around Cairo University. It looks like a little metropolis. The security measures were pretty noticeable, some iron fencing, and so on. Syed told me there had been some disturbances. But the university has since 'taken care of it.' I guess that means the university had to shut down the political student groups on campus. The old democracy versus security issue, I guess. I didn't get a chance to see the sights so much but we did drive around a bit. It's just as well. Now that you're here, we can check things out together after we figure where we're going with this Khaled character."

"What do they eat here?" she asks. "I mean, what kind of foods do they like?"

Linda must be hungry by now. She probably hasn't eaten much of anything since earlier yesterday. "Syed made me some breakfast they call 'ful.' It's pretty good. He likes to drink a lot of tea and Turkish coffee. They seem to eat bread with every meal. Oh, and he gave me this dessert. I forget what it's called. It was delicious. Of course, everything tastes good when you're hungry. I'll see if I still like it if I have it again. Their coffee is very strong; they serve it with the grinds floating on the top. Syed says to let the grinds settle to the bottom before drinking. I'm still getting used to their customs and waiting for the moment when I inadvertently insult our host by doing or saying something stupid; which will probably happen soon if it hasn't already. Syed is the perfect gentleman. He's no doubt already overlooked several of my faux pas."

"He seems very nice," says Linda. "I noticed that when I ask him something, he tends to look more at you when answering. Another Egyptian mannerism, I suppose. Maybe he feels awkward around an American woman, I don't know. I'll get into the swing of things soon enough. At least we've both seen enough movies to know to take our shoes off before entering an Egyptian home."

Linda wants to follow up on our earlier meeting with Khaled and starts to ask me about the plate and what I thought it might contain when Syed comes back inside.

"Everything all set?" I ask.

"Yes, all is fine. We should have another piece of our puzzle shortly." Seeing that both of our glasses are now empty Syed asks if we would like another drink and, in typical fashion, takes off to collect another round without waiting for a response.

I look at Linda. "Well, we really shouldn't refuse our host, don't you think?"

Linda smiles, managing a polite attempt at laughter. And then, getting into the spirit of the evening, she says "Well, I guess if I can't have my glass of wine, I'll just have to settle for volume to make up the difference. It's quite good."

Syed returns with our drinks; one in tow for himself. After he sits down Linda surprises us with an announcement.

Looking at neither of us in particular, Linda tells us she had done some "minimal research" regarding what we have been calling "gaps" in the tablet.

"There is a new light detector that a research group at Boston University is working on," she says. "It is a very thin, graphene-based prototype that can absorb light to see terahertz waves. It used to take very cold temperatures for these types of devices to work properly, but this new prototype can operate at room temperature. The upshot is that it can pick up anomalies beneath the surface of even opaque materials. There is more to it than this but the bottom line is the device can generate more electrons, creating an electric signal that might help us decipher whether or not there is any additional information hidden in the tablet. Maybe whoever produced the tablet embedded additional information inside, just out of reach of most modern scanning devices. From what I have been able to figure, it seems that its intended use is more in the medical and security fields than for our needs. But I don't see any reason, at least in theory, why it couldn't be used to help us in our investigation."

Linda, ever true to form, has once again shed some light on our problem and offered a plausible solution. If this device can do what Linda thinks it can, it just might reveal another layer of information to help us in our ongoing attempts to solve this mystery. My money is on Linda. Either way, it would certainly be worth pursuing.

"This device—this prototype—do you think it is developed enough to be of use to us?" I ask. "Would they let us have access to it?" I'm sure Linda had considered these possibilities but I want to know what she thinks our chances are of pulling this off. I am also worried about revealing our secret to a bunch of strangers—albeit scientists—who might have more than a few questions for us.

Linda pauses for a moment . . . lost in thought, wide-eyed and staring off into space. "We'll just have to figure it out," she says.

This is a first. Linda hadn't worked her way through this particular scenario. I feel like pouncing on a rare *gotcha* moment but think better of it and decide to let her off the hook. It wouldn't be fair. After a twenty-hour flight, little food, and a bottle of beer, Linda is still pretty sharp as far as I'm concerned.

Syed has been sitting quietly taking everything in. "This sounds like a very useful instrument," he says. "The pure form of carbon should hold its energy level much better. There are electrodes to produce what you call the signal; to provide an imbalance—"

"You're a quick study," says Linda, obviously impressed at Syed's ability to grasp her explanation so quickly.

"So, I see you have something in common with Dr. Banner," he says.

"What do you mean?" Linda asks.

"Jake is full of your American expressions. He is constantly testing me to see if I can understand what he is saying."

Linda looks a bit nervous. "Oh, I'm sorry. I didn't mean—"

"No, it is quite alright." Syed does not appear upset. "It is a game we play with each other. Sometimes I even . . . stump him up. I was a 'quick study' when acquiring my education. Jake keeps me 'on my toes.' "

I am delighted Syed and Linda are getting along so well. It can only strengthen our collaboration.

"Let me bring you something to eat. You must be *famished*." Syed is once again showing off his growing command of the English language.

"Oh, that's alright, I'll—"

"You'll get used to it," I say, looking at Linda as she watches Syed walk away. "He only asks if you want anything to be polite; you have no choice in the matter."

Syed returns with what he calls "Halawa," a middle-eastern concoction consisting of various nuts and almonds in a sesame paste.

Linda thanks Syed and consumes the meal with what I would call "subdued vigor." She is hungry and the beer hasn't exactly provided the nutrients she needs; although, there is something to be said for the barley and hops.

After Linda finishes the Halawa, she thanks Syed once again for his hospitality, excuses herself and retires for the evening. She is—to use an American expression Syed would appreciate—spent.

I might as well ask Syed straight out about what he thinks of Linda. Who knows, maybe his good-humored repartee with her was just for show. As I understand Egyptian custom, Egyptian men aren't used to broaching the subject of women except, perhaps, in the general context of enquiring about one's family. I better leave it alone.

As Syed and I finish our lagers I figure now is a good time to bring up one of the illustrious alumni of Cairo University, Omar Sharif.

Syed is in his element once again. After mentioning, with noticeable pride, that Omar Sharif had been born in Alexandria in nineteen hundred and thirty-two, he proceeds to give me a rundown of Omar's career. Starting with Sharif's film debut in *The Blazing Sky* and ending sometime later with such films as *Heaven Before I Die* and *Return of the Thief of Baghdad*, Syed regales me with his cinematic expertise. He knows the original names of Sharif's movies as well as their popular titles. In addition to sharing the details of Sharif's career, he also provides many technical aspects of film production. He speaks at some length about cross-cutting; aerial, bridging and tilt shots; and a host of other techniques that are all Greek to me. He gestures emphatically as he describes an especially dramatic scene, explaining in elaborate detail how it was choreographed.

It is already getting late by the time Syed finishes his lecture on cinematography and we call it a night.

My first impression of Syed was that he was an older, settled gentleman. As I have gotten to know him better, I can see more of the young man he once was. His enthusiasm; his boyish charm; and his unflinching determination to arrive at the truth reminds me of myself in some ways.

Why do we start out seeking, yearning, fearlessly challenging the unknown, only to "cash in our chips" and give up? What happens to us? Why do we lose our zeal? Why do we . . . settle?

According to the American author and biologist Dr. Paul Ehrlich, ". . . we are creatures of accident clinging to a ball of mud hurtling aimlessly through space." . . . That's it? No deeper meaning? No hidden secrets that might eventually reveal a grander scheme—a larger purpose for us to discover? I'm sorry; it's just too *"cut and dried"* to my liking.

Richard (Ritty) Feynman, a theoretical physicist, attempted—unsuccessfully in my opinion—to close the door on any further angst when he wrote:

> I have approximate answers and possible beliefs in different degrees of certainty about different things, but I'm not absolutely sure of anything, and of many things, I don't know anything about, but I don't have to know an answer. I don't feel frightened by not knowing things, by being lost in the mysterious universe without having any purpose which is the way it really is as far as I can tell.

Okay, where's the cyanide? I never got a chance to speak with Ritty, although I did meet him once late in his life. It was one of those, "Hi, how are you? greetings," that didn't allow any time to delve into complex philosophical issues. Had I an opportunity to press him on this statement, it might have gone something like this:

JB: So, what, exactly, do you mean when you say you 'don't have to know an answer'?
RF: You're reading too much into it Jake; take it at face value. It's quite simple: There are some things I know and some things I don't. The things I don't know, I don't have to know.
JB: But there must have been a time when you felt you did 'have to know' . . . right?
RF: Sure, I sought the answers to the deeper questions like everybody else. I went through the normal process: Seek—no answer—rinse and repeat. After a while, you just chuck all the philosophical baggage and decide to 'live in the moment,' as they say.
JB: So, you agree with Carl Sagan that we are orphans in the universe with no meaningful explanation for our existence and no cosmic or divine plan?
RF: Close enough.
JB: When did you stop asking, stop seeking, stop yearning for an answer to life's meaning?
RF: That's a good question. I can't say I know . . . honestly. It just kind of slipped away. I mean, there wasn't any epiphany moment; nothing like that. I just stopped thinking about it. Not permanently, it comes and goes; but I don't dwell on it anymore.
JB: Thanks!

What else can you say? The pattern is repeated all over the world—day in and day out. People give up; it's easier. With all due respect to the late Richard Feynman, I'm not ready to throw in the towel . . . just yet.

Chapter Nine: More Secrets

Syed is enjoying his morning tea while I try the Turkish coffee when we are joined by Linda who had slept in late. I had told Syed earlier that Linda and I are in the habit of having a light breakfast or skipping it altogether, usually settling for coffee and juice. He suggests we have something to eat in the city after we meet with Khaled, to which we readily agree.

"So, don't keep me in suspense," says Linda. "When am I going to get to see the tablet?"

Just as I had originally believed, Linda must think the tablet is sequestered away in a safe somewhere; perhaps at the university.

"I would be glad to show you the specimen," says Syed, matter-of-factly as he gets up to retrieve the artifact.

As soon as Syed leaves the room Linda gives me the look: Eyebrows launched upwards, eyes staring right through me, mouth slightly open. This is her personalized version of the "What the hell were you thinking?" look. She should patent it.

Linda has no intention of letting me off the hook. "Syed, had the tablet here all this time?"

"Yes, but I didn't think—"

"Never mind Jake," she says, "Let's just stay focused on the task at hand."

This would be a good time to say nothing. Silence is golden.

Syed brings the tablet into the living room and carefully lays it down on the table in front of the sofa where Linda and I are sitting. He then sits on the sofa opposite.

"This is amazing!" says Linda. "I don't see any markings on the tablet. How did these symbols get embedded in the stone?"

Linda has either forgiven or forgotten about my omission in not telling her about the tablet earlier. She zeroes in on the most important feature—the remarkable inscription appearing in the stone without evidence of tool-work. "This is very strange," she says as she stares intently at the tablet. "There doesn't seem to be any way to know for sure how the inscription was made."

Emile rings the doorbell and Syed takes the cue: "Perhaps we should get going," he says. That is all the invitation Linda and I need.

It is a little after nine in the morning when we arrive at the Citadel which has just opened for business. Syed explains that the area surrounding the ancient walls of the Citadel is usually closed to traffic to keep the attraction secure for tourists and visitors. Emile manages to get us within proximity to the main entrance. Syed gives Emile instructions and we begin a leisurely walk around the grounds, taking in the view. Syed tells us the Citadel is a popular tourist landmark and points out some of the more salient features. There are three mosques visible from the esplanade, the Mohammed Ali Mosque, in particular, caught our attention. It is quite large and has several prominent, domed edifices on the front portion of the mosque. The air is uncharacteristically clear and we can just glimpse the pyramids off in the distance. It is like a pleasant oasis in an otherwise rambunctious, teeming metropolis.

Syed spots Khaled off in the distance carrying what looks like a shopping bag. Khaled notices us and begins walking toward us. We find a bench to sit on and Khaled surreptitiously removes the tablet from the bag. We now have a second piece to our puzzle.

Simplicity itself; just like the inscription on the first tablet—no embellishments, no complex details; just the bare essentials: An image of a double-stranded polymer molecule, DNA.

"DNA?" says Linda, clearly as perplexed as I was when I saw the first tablet.

Syed pays Khaled the five thousand EGP as previously agreed. We assume that will be the end of our brief association with Khaled as he had already balked at revealing any additional details about the location of the find or any particulars about his accomplices. He has another surprise up his sleeve, however. He says there is one more tablet that he knows of but doesn't have access to it. He tells Syed that he and his cohorts had parted ways after their latest escapade had gone sour. Syed doesn't bother asking him why he is only now bringing up this important information. Instead, he asks Khaled what he can tell us about the remaining tablet. Khaled takes a cell phone out of his pocket and shows us a picture of the third and final link to solving our mystery. It is a picture of the image on the remaining tablet: A representation of the cosmic microwave background radiation, the thermal radiation left over from the Big Bang.

Speaking in broken English Khaled says "I did not think you . . . would be . . . interested." Khaled is once again displaying his ignorance in scientific matters.

Linda and I are dumbfounded! We both stare at the picture in disbelief. Unless the picture has been photo-shopped it is an image of the cosmic background radiation from the early formation of the universe. And it is in color! This is just plain weird!

"It looks out of focus like the image is shifted or I'm seeing double images of the background." Says Linda. "Do you see it?"

"It could be a distortion in the image caused by the camera angle or reflected light. We'll have to wait until we can take a look at the tablet up close; that might give us a better idea."

Syed speaks a while longer with Khaled in Arabic and has him transmit the photo to his cell phone. He then takes the tablet, placing it back into the bag, and ends the conversation. We watch in silence as Khaled's small frame quickly shrinks into the distance.

"We have much to discuss," says Syed. "How about something to eat?"

Syed is certainly taking things in stride. It is difficult for me to predict his response in any given situation—he can remain stoically calm under stress and surprisingly focused when others would easily lose their concentration.

"I'd like that," I say. "Linda, you must be hungry."

"I could stand a bite to eat," she says. She then looks at me as if she had just remembered something. "Oh, by the way, we should kick in our share of Khaled's payment. You're the mathematician—how much is one-third of five thousand?"

"Egyptian Pounds," I say. "That would be about—"

"Do not bother," says Syed. "It is not a large sum. I can easily afford it. You have already paid a fine price in getting here to help me with this mystery. It is I who owe you."

I have learned not to argue with Syed's generosity and Linda follows my lead.

Syed has Emile take us to a restaurant nearby that caters to Western tastes. Syed speaks to Emile for a moment and then dismisses him. We order a typical American breakfast. After ordering, Linda brings up the topic that has been on all of our minds. "How do we get our hands on this other tablet?" she asks.

Syed is careful in answering so as not to raise our hopes beyond realistic expectations. "Khaled Mosi will make an effort to reach his former colleagues. He will contact me as soon as he has something to report. I wish I could be more positive but I do not have much confidence in this Mosi and his gang. We will see."

I appreciate Syed's candor and insight. His judgment is spot on, as usual. "How soon do you expect to hear from him?" I ask, hoping to narrow the time frame.

Syed pauses for a moment, looks off into the distance, and then, reengaging, says "I think I will hear from him soon enough. His kind can track money like a bloodhound."

Linda leans back in her chair and reflexively begins to cross her legs. Remembering her brief tutor on Egyptian customs, she reverses herself quickly and placed her feet firmly on the floor. "I'm curious about the coloring process on the stone showing the cosmic background radiation. It looks like it could be a normal lithographic printing but it's hard to tell from the picture."

"I'd like to get my hands on it," I say. "By the way, I wonder if this new tablet has the same 'gaps' as the original? That might be telling. What do you think, Syed? Could we do another analysis without giving too much away?"

"I will look into it. I do not know if I can repeat my former success, but I will give it 'the old college try.' It would be very interesting if we would discover the same 'gaps.' And I think you are right; it may tell us something we do not already know."

The waiter comes over with our order on a moving food cart and places our food and drinks. He keeps looking at Linda, trying not to be too conspicuous but failing in the attempt.

I suggest to Linda that she try the Turkish coffee and she reluctantly agrees. While we are talking our coffees have time to settle and Linda takes a sip. "It's pretty strong," she says.

That's an understatement but I say nothing. "When do you think we will be able to get these tablets examined with the new scanner?"

"I'd like to get it done as soon as possible," says Linda. "As I see it, we have two obstacles to overcome. The first is access. We need to not only get permission to use the equipment, but we also need at least one of the researchers to help us use the equipment. And the second thing is confidentiality. We may have no choice but to trust one of the technicians with our find."

Syed has been ignoring his glass of coffee, following the conversation. He slowly takes a sip, pausing for a moment. "It may be necessary to make a personal visit to see these people in person—do you think?"

"That's a good point," says Linda. "It could be risky to seek their help without the personal touch. It will take some . . . salesmanship." She looks at me expectantly.

"What do you think?" I ask, deferring the matter to Syed. It is his find after all.

"I should know about the possibility of acquiring the additional tablet very soon. Once we have it secure, we can make plans to analyze these 'gaps' and see what we may find. Meanwhile, I have neglected to show you our fair city. You, of course, will want to see the pyramids and so on. Let me make a few calls and see what I can arrange."

Linda and I are both excited about the prospect of seeing the pyramids up close and otherwise taking advantage of our visit. "That sounds like a great idea!" I say. "What do you think Linda? Are you up for it?"

Responding to Syed, she says: "Thank you so much; I'd love to see the pyramids—"

"That would be the place to start, but of course there is so much more to Cairo than the pyramids. Perhaps you 'kids' would like to see a little of the nightlife as well?"

Other than solving our mystery, there is nothing more I would rather do than share the evening with Linda—and our host, of course. "That would be terrific Syed, but you have already—"

"I will not hear of it further. Anyone who visits Cairo, especially for the first time, must see the pyramids and the various attractions. What is your saying? 'All work and no play makes Jack a dull boy.' Or should I say, Jake?"

There is no ignoring or refusing Syed's hospitality. "That is very kind of you Syed. We are looking forward to it."

Syed pushes his chair back and stands up for a moment. He then reaches for his phone and excuses himself. "I will be just a moment," he says.

Linda and I finish our breakfast and take a moment to compare notes.

"If we can wrap things up in the next day or so," I say, "we'll still have to figure out how we're going to get access to this new scanner; if that's even possible."

Linda seems more relaxed. She probably feels like she needs to be on her best behavior around Syed. Neither of us wants to jeopardize our relationship with him. He is, after all, the one who brought us into the loop. It was Syed that had come into possession of the original tablet. We are both fortunate to be included in the search for answers to this mystery.

"Let's not worry about that now," says Linda. "I'm looking forward to doing a little sightseeing. I can't wait to see the pyramids up close. Has Dr. Azad ever talked to you about any of his work in archaeology?"

"Yes, he has regaled me with a story or two. He has done some work in Luxor. It's just south of here. I think he called the location *The Valley of Nobles*. He didn't get into it very much. If you want to grab his attention just start talking to him about old classic American movies. He loves the subject."

"That's funny," says Linda. "Who would have thunk it? What's his favorite?"

"I don't know; he seems to like most everything. He is particularly fond of *The Bridge on the River Kwai*. We had a deep philosophical discussion about the meaning of life—"

"It was about that scene on the bridge with Alec Guinness and the Japanese commander," says Linda. "I loved that scene!"

I can't help laughing. "Just bring it up and you'll be on his good side for sure."

"Thanks for the tip."

Just then Syed returns and sits down. "Emile will be here shortly," he says.

Seeing the Great Pyramids up close with Dr. Azad as a guide is priceless. Syed goes beyond the usual folderol one would expect from a paid guide: How many workers were involved? How long did it take to build? Its symbolic representation of the sun's rays. How the sides of the pyramids assisted the Pharaoh's soul in reaching the gods above and so on.

Syed explains that the Great Pyramid was built for the Pharaoh Khufu. He is very candid when speaking about the interior construction of the pyramid, offering no definitive explanation for Khufu's instruction to build such a complicated system of passageways and interior chambers. He expands at some length on the differences in construction techniques involved in creating tunnels and vaults above and below ground. He also explains that the Great Pyramid was the only one of several dozen constructed during the period (2600-1750 B.C.E.) having such tunnels and vaults *above* ground.

Other Egyptologists have speculated as to the reasons for such elaborate structures within the chambers of the Great Pyramid, some suggesting that the design evolved during the Pharaoh's reign to reflect his increasing divinity during his rule. Syed offers no such explanation, attributing such features to the whims or caprice of a divine king. "No one knows for sure," he tells us.

Syed begins to discuss the differences in the volumetric arrangement of the many vaults and passages within the interior—the geometric and mathematical features of the pyramid—but leaves off after a brief review, thinking it best to defer for another time. Syed knows this esoteric subject is dear to my heart. He also knows that we had pretty much vetted such arguments as far as we could and doesn't feel the need to expound further.

Linda seems to be enjoying the tour, leaving the details to our expert host. The weather is cooperating and allows a modicum of comfort. Wearing pants with a long-sleeved top and wedge sandals, Linda seems comfortable in her attire.

"Do we know for sure who built the pyramids?" Linda asks.

"Yes," says Syed. "Extraterrestrials built them."

Linda and I laugh. I'm happy to see that Syed is comfortable enough around Linda to make such a jest.

Linda, feeling much more comfortable around Syed as well, then asks "How did they manage to lift such heavy blocks?"

Syed, assuming a more serious air, replies: "There have been many proposals offered to explain how these huge blocks were lifted to such heights but the controversy remains. Many Egyptologists no longer believe the pyramids were built by *slave* labor. The laborers who worked on the pyramids were provided for and lived in what has been called 'pyramid cities.' The blocks were too heavy to lift by hand and, as Dr. Banner can easily explain, such an attempt would have been impractical. No, some other method had to be used to accomplish the work.

"The process is not nearly as complicated as some would have you believe," he continues. "I am in favor of the proposition that has been put forward by Professor Redford. He is the instructor of Classics and Ancient Mediterranean Studies at Penn State University. The story he advances is quite simple: After the stones were quarried there would be teams of oxen and many laborers to haul the stones along a ramp which has been lubricated to ease their movement to the construction site. They would then build the ramps to place the stones, using the ramps to lift them as they went along. They of course would have used ropes and a pulley or winch of some kind to give them the necessary leverage but nothing more complicated than that."

We are unable to challenge such a proposition and have no intention of doing so.

Linda is now getting into the spirit of things and beginning to appreciate the font of knowledge that Syed has at his disposal. "How long did it take to build all of this?" she asks. "It's enormous."

Syed is now in full professor mode. "The best estimates," he says, "place the construction period close to twenty years or so. Some believe it was much longer. This whole area was once a great necropolis and there were a great many temples and smaller pyramids. The entire complex was once surrounded by a much larger enclosure known as the *Wall of Crows*. The Khafre and Menkaure pyramids you see were added sometime later and together we refer to the site as the 'Great Pyramid Complex.' "

Syed is now hitting his stride as he continues to entertain us with his knowledge and showmanship. His enthusiasm is contagious and we are both captivated by his impassioned delivery. During one particular exposition, he thrusts his right hand outward in a sweeping gesture expounding on several fascinating features of the pyramids. He then takes us to *the pièce de résistance*: The Great Sphinx of Giza—the oldest sculpture mankind has ever known.

"We do not know who built the Sphinx. Some say it was authorized by Khufu or his son, Khapre." Syed is careful to avoid speculation. "The sphinx may have at one time been quite colorful and there are still some small remaining touches of paint you can see near the ear if you look closely. We do our best to keep it in a good state of repair; the restoration work is ongoing."

"What is the sphinx supposed to represent?" Linda asks.

"The body of a lion and the head of a god, or king, is said to be a symbol of strength and wisdom. Interestingly, it was carved from the foundation of the Giza plateau."

With a sudden flourish, Syed then exclaims "Even though the desert sand has buried the sphinx many times in the past, yet it lives!" And then, on a more reflective yet still humorous vein: "It is a little worn as you can see but I should be happy to look so good after well over four millennia!"

We all have a good-natured laugh, ending our tour on a high note.

Over these last few days, I have reflected upon my relationship with Syed. I have come to realize that what had begun as a formal, professional liaison has progressed to one where we are now comfortable sharing our deeper concerns and aspirations. He expresses no qualms regarding the requirement of prayer five times daily—at dawn, noon, late afternoon, sunset, and night—and engages in this holy obligation freely and without reservation whether anyone, including me, is present. His religious convictions are on full display, his friendship with me unreserved, and his commitment to an honest search for the truth unchallenged. He is, in a time-tested phrase, his own man . . . and I am proud to know him.

Chapter Ten: Interlude

A great confluence of time and circumstance has brought Syed, Linda, and me together. To what end? None of us know. We have nothing but speculation and suspicion to guide us. I wonder if we have discovered anything new at all. The two, possibly three, plates tell us nothing we did not already know. Schrödinger's equation, the DNA molecule, and cosmic background radiation are known features of the world. We had learned nothing new about the universe and our place in it. For all we know, these tablets are the "be all and end all" of our grand and noble quest. Perhaps there is nothing more to discover. Have we come so far only to end our search empty-handed? Well, if not empty-handed, perhaps short of what I seek: A fuller, more complete explanation of the origin and evolution of the universe; a deeper understanding of life and our indispensable role in the outworking of some cosmic plan; and the final, complete mathematical formulation that would tie it all together.

Here I am, presented with an astounding discovery . . . asking for more.

Part of the pleasure of eating a good meal is the *satiation* of hunger. It is when our appetite has been satisfied and we keep eating that we run into trouble, adding unwanted and unnecessary poundage. If my "appetite" for knowledge was substituted for food I would weigh several hundred pounds. I have always had to know the answers to the hard questions and am never satisfied with anything less than a complete understanding of how things work.

Back in junior high school when I took my first algebra class, I learned to manipulate the variables with considerable ease—so much so that, when I sat for final exams, I used *ink* instead of a number two pencil. I aced it! There was only one problem: I didn't understand what I was doing; I just knew how to "do the math." It wasn't until sometime later that I was hit by an epiphany. I don't remember the particular circumstances. All I remember is having a real-life math problem where I had but two pieces of a puzzle and needed a third, derived from the other two. It was at that moment that the proverbial "light bulb" went off in my head and I finally *got it*. Ever since that moment I made a personal vow —not just to learn—but *understand*. And, as Robert Frost relates in *The Road Not Taken*, it has made "all the difference."

Linda and I are looking forward to this evening. She has taken some time to rest, recovering from her "sleep deprivation."

Syed excuses himself from this evening's festivities. He says it is nothing to be concerned about, he just isn't feeling up to it. He hadn't been feeling well since our return from the necropolis. I don't make anything of it and tell him he should get some rest. He insists Linda and I go ahead with our plans.

Linda comes into the living room dressed in a long black gown with a high neckline and a matching top with long sleeves. I tell her the news about Syed.

"Oh, that's too bad," she says. "I hope he's alright. Well . . . no matter; we'll just stay in. We probably shouldn't go out with Dr. Azad feeling—"

"No, he left instructions for Emile to take us to a *swank* nightclub. Everything's all set."

"Are you sure?"

"Yes . . . Syed will be fine. Perhaps 'clubbing' isn't his thing. He once mentioned that he had been to a club before but it may have been on business. Who knows? He insisted we go ahead and you know by now not to refuse—"

"Yeah, I guess so. We're grownups. We don't need a chaperone . . . do we?"

I can't tell if Linda is teasing or serious. "We're technically tourists," I say, hoping to allay any concerns she might have. "We're exempt!"

Linda laughs triggering me to join in.

After a few moments, Emile arrives.

"Our carriage awaits us!" I say enthusiastically.

Emile takes us to the Club Absolute. When I ask him to describe the club, he gives a one-word reply: "Expensive."

"Yes, but what about the atmosphere? What's it like?"

Emile's dismissive tone reveals his socioeconomic attitude toward such establishments. "There are many 'ex-pats' who go there, along with the wealthy elite. You should like it just fine. It has all of the Western decadence that tourists prefer."

I feel like telling Emile: "So why don't you tell us what you really think?" but decide to drop it. I hadn't any idea what Emile's attitude is about pretty much anything. This is an eye-opener for me. At least I know where he is coming from.

Lifting my eyebrows, I look at Linda, my expression silently communicating my muted surprise at Emile's take on western civilization. She just shrugs her shoulders, displaying a casual indifference; Emile's comments coming as no surprise to her.

Human nature is a funny thing—individual life experiences producing such disparate outcomes. Where Linda sees an unremarkable example of resentment or animosity toward the privileged, I need to understand *why* Emile feels as he does. Given the same circumstances and another chauffeur—say another Frenchman named Pierre with another life experience—I can imagine a much different outcome. Pierre might have extolled the "fantastic ambiance" and "international appeal" of one of the "finest" clubs in Cairo. Are we simply the product of our experiences, having little or no choice in how our prejudices and opinions are shaped? Could Emile, in a moment of deeper reflection, reverse his attitude?

We arrive at the club and Emile tells us to give our names if asked. "There should not be any problem," he says. "You have my number. Just give me a call when you are ready to leave."

Walking toward the club I ask Linda if she has that déjà vu feeling. It is obvious we have arrived back at square one: We had been here yesterday when we met Khaled. The Club Absolute, we realize, is just the other half of the Casino El Shagara. We are now in the lower level. The Sangria where we met with Khaled is on the upper level.

"This must be the center of the universe in Cairo," says Linda.

"Must be."

We have no trouble getting inside. Our Western appearance, despite Linda's effort to blend in as best she can, apparently carried sufficient cachet for the black-clothed security guard to wave us through.

I can see the attractive draw of the club as soon as we enter. It has a certain atmosphere that might appeal to the upper echelons of society, I suppose. As we walk over to the dance floor area to have a look-see, I notice the now familiar, yet still enchanting view, of the Nile. The décor is striking, the music an eclectic European-Arab mix, and the ambiance surreal. The clientele is a mix of tourists, ex-pats, and some locals (as far as I can tell), all looking quite at home; giving the impression that they are enjoying just another casual night out rubbing elbows with their fellow elites.

The bar is busy but we prefer a table anyway so we wander through the crowd until we spot a table where a couple is just about to leave. The gentleman sees us approaching and waives us over. I thank him and we sit down. As I look around, the music pounding out a relentless staccato beat, I see a constant stream of motion—people scampering through the club going from one end to the other. No sooner have we sat down than the couple next to us get up and scurry away.

"I feel like we're watching a version of *musical chairs*," I say.

"It's got all the hustle and bustle you could want from big city life," says Linda. "Shall we order a drink?"

"By all means," I say. "Your usual?"

"Here I am halfway around the world in an exotic foreign land and all I can think of to drink is my *usual*. Maybe I should try something different. What's that drink you like?"

"It's called a 'Sidecar.' "

"What do they put in it?"

"I've been drinking it so long I almost forget. Some cognac, orange liqueur and I don't remember the rest. Don't forget about the ice . . . you know."

"Oh yeah, that's right. Well, no sense *tempting fate*. I guess I'll have a Cabernet . . . my *usual*."

"You ever wonder where all the old expressions come from," I say, attempting to kick-start a conversation. "Many of them come from maritime terminology. Toeing the line for example refers to the way the Captain would have the crew line up on deck. The crew members would place the toes of their shoes on the planks to form a straight line."

"What if the planks are crooked?" she says.

We have a good laugh which *breaks the ice.*

"And what's that song? Because you're mine, I *walk the line.* What's that all about?" says Linda, taking my queue.

"Something about being faithful, I think. You know, playing it straight; not veering off into infidelity."

"It could be referring to a DUI and the guy's trying to walk the line to get off on a drunken-driving charge so he doesn't get thrown into the tank and his wife has to come and bail him out!"

We have another good laugh, Linda enjoying our little repartee.

"I think I will try the wine," I say. "What would you recommend?"

"What do you like?"

"I've tried a Merlot and Beaujolai—"

"You can forget the Beaujolai," she says. "To use your maritime vernacular, 'that ship has sailed.' You could order it but it passed its peak several months ago. Are you up to trying something a little different?"

Linda's derisive tone leaves no doubt in my mind that she thinks Beaujolai is for sissies. "Sure," I say.

"Well, there's a wine called Syrah or Shiraz that might work," she says. "I think you will like it."

Keeping the playful spirit of the evening going, I lift my right arm and—elbow firmly planted on the table and my palm extended outward—say, in my most commanding voice: "Make it so!" I think evoking the image of Jean-Luc Picard, commander of the starship *Enterprise,* is a nice touch.

" 'O, Captain! My Captain!' " Linda responds in a good-natured rejoinder.

"Now you're quoting Walt Whitman!" I say. "I'm impressed! Robin Williams played—"

"I know," says Linda *"Dead Poet's Society . . ."*

107

"How does the expression go: 'You're not going to believe this but' . . . I met Robin Williams at a baseball game in San Francisco."

"You at a baseball game?" says Linda, calling me out. "I'm sorry but I'm not buying it."

"Why not? I may not be crazy about hot dogs but I like apple pie; why not baseball?"

"Okay, tell me what *the designated hitter rule* means."

"That's easy," I say, bluffing. "It designates when the batter can hit the ball."

Linda laughs so loud she attracts the attention of several onlookers.

"Alright smarty-pants, you tell me what it means . . . *without* looking it up on Google."

"It is a rule adopted by the American League that allows a team to designate one player to hit in place of the pitcher."

"How'd you know that?"

"I have four siblings and they're all boys; do the math."

"Jeremy had two tickets and asked me to go with him," I continue. "He said it was a special game between the Giants and Yankees. The two teams don't usually play against each other during the year he said."

"Oh yeah, that's one of those . . . never mind. Who's Jeremy?"

"I don't think you know him. He teaches astrophysics. He was working on some interdisciplinary courses and wanted my take on some ideas he had. We went to the game some time ago."

"Did you get good seats?"

"Jeremy seemed to think so—"

"Where did you sit?"

"Somewhere behind first base, about six rows back or so. Robin was sitting a couple of rows behind us."

"You had better seats than Robin Williams? Cool!"

Did Linda just say "cool"? "I didn't say anything until the *seventh-inning stretch*," I say, pausing long enough for the baseball terminology to sink in. "Aren't you impressed I know what that means?"

"No; everybody knows what that means."

Letting her glib remark pass without comment, I continue: "I left my seat to take care of business and on my way back I figured 'nothing from nothing leaves nothing,' so I went over and introduced myself."

"What did you say? What was he like?" she asks.

"I don't remember what I said when we first met, but I do remember his reaction when I walked over to him."

"Was he upset?"

"No, just the opposite. He immediately got up from his seat and gave me a firm handshake."

"What did he say? What did you talk about?"

Linda is listening intently. My little vignette with Robin Williams is holding her attention.

"I brought up a show that had just aired where some would-be actors and comedians did impressions of *real* celebrities. I had only watched one show. It was probably canceled shortly after airing. I told Robin about a guy on the show that did an impersonation of him and that I thought it was pretty good. He asked me a few questions about the show but I got the impression he hadn't heard of it and didn't know what I was talking about. The whole time he was a perfect gentleman. Jeremy said Robin and Billy were rooting against each other's teams."

"Billy"? Linda says. "Billy Crystal was there too?"

"Oh, yeah . . . he was sitting next to Robin," I say matter-of-factly.

"That makes sense," says Linda, "East and West Coast rivalries—Billy for the Yanks and Robin for the Giants. Classic!"

"I guess."

"What was Billy like?"

"I don't know. He didn't say anything that I remember. I was too distracted talking to Robin to pay much attention. Billy was probably used to Robin getting most of the attention. During the game, Robin would shout something out and everyone would laugh. The game went extra innings and I remember Robin shouting: "Time for the eleventh inning stretch!"

"It's pretty sad the way he went," says Linda.

Not wanting to go down that path, I change the subject. "Do you like it here?"

"Not really," says Linda. "I'm sure it appeals to a certain set but . . . I don't know, it just seems too—what's that old expression—'yuppie' for me. I guess we're just a couple of old fuddy-duddies."

"If we are fuddy-duddies, you're the youngest fuddy-duddy I know."

"There's another expression—"

"We could go upstairs to the Sangria. They have a terrace where we could sit . . ."

"Okay," says Linda, and off we go.

We find a nice spot upstairs on the terrace of the Sangria. The music is upbeat but not very loud. Our table is near the corner of the terrace, sufficiently sequestered to allow conversation without too much distraction. Our strategic location also allows us to escape the occasional dancing duet that, not having an *official* dance floor, make their own by dancing between the tables. We take what we can get.

After a long wait, exercising considerable patience, a server finally comes by and takes our order. The staff doesn't seem very motivated.

When our drinks arrive, I notice my large glass is slightly cool to the touch. "I thought the wine was supposed to be served at room temperature," I say, offering what little knowledge I have.

"It *is* room temperature," says Linda., "in France."

"Point taken," I say. What do I know about wine anyway? My only claim to being a wine aficionado was an aborted effort to take an online class. Even if I had finished the course and been graded on the curve, I wouldn't have deserved a grade higher than an F-plus.

"I saw a *Starbucks* the other day when Syed and I were out and about," I say, trying to recapture our earlier mood.

"Oh yeah?"

"Cairo reminds me of a building project where the original architect's master plan was going along smoothly and then something happened and the project was abandoned. Then some other people came in with different ideas and took over. Western businesses are scattered about the city as if they had been airdropped in by the enemy."

"Everybody needs *Starbucks*," she says.

"Fair enough," I reply, "but do they need *Kentucky Fried Chicken* and *Burger King* and *McDonald's*? Western enterprise has permeated the planet. It would be nice if there was a spot left on the Earth that was uncontaminated by—"

"What does that even mean?" says Linda. "*Uncontaminated*"? Everything is contaminated—there is no pristine spot on the Earth where civilization has not encroached and infected the planet. Every nook and cranny has been explored, exploited, and exterminated."

" 'What a gift for alliteration you have! I dare say I could get you a job in a shop, maybe—' "

"Oh, now I'm Eliza Doolittle, am I?"

"No, just the opposite; you're the finished project."

"I bet you men enjoyed that movie. What's not to like? A horny, middle-aged man—"

"Who says Professor Higgins was horny?"

"All middle-aged men are horny."

"And how might you know?" I ask, half curious and half hopeful.

"The whole premise of the movie . . . or the play was how a wealthy middle-aged man could get his jollies by having his own personal, live doll to play with."

"Well there is that," I say. "But you must admit it was a noble effort on Professor Higgins' part to lift a 'gutter-snipe' from the depths of poverty and transform her into the social envy of the privileged class. I thought that was the *overarching narrative* as a critic might say."

"Did you know that, in the original play, *Pygmalion,* the professor doesn't get the girl?"

"That was stupid," I say. "I'm glad they changed it. I liked the part where Professor Higgins is extolling the virtues of men—"

" 'Why can't a woman be more like a man?' "

"Exactly," I say. " 'Men are so pleasant, so easy to please; wherever you're with them, you're always at ease—' "

"You are pleasant Jake, and I am always at ease 'wherever' I'm with you. Perhaps there are *some* men where that part of the lyrics is true."

This is the first time Linda has ever said anything approaching a hint that she might like me. If she had minded being around me, she wouldn't have accepted my invitations for dinner the few times we went out together. But still . . . this is nothing short of a passionate announcement of her intentions towards me as far as I am concerned.

Linda's comment reminds me of the scene in *Dumb & Dumber* between Mary and Lloyd when Lloyd, played by Jim Carrey, asks Mary, Lauren Holly, about his chances of capturing her heart:

Lloyd: Hit me with it! Just give it to me straight! I came a long way just to see you, the least you can do is level with me. What are my chances?
Mary: Not good.
Lloyd: You mean, not good like one out of a hundred?
Mary: I'd say more like one out of a million.
(*pause*)
Lloyd: So, you're telling me there's a chance... YEAH!

I bet I can beat these odds, or at least narrow them down a little. "What would you like to eat?" I say, getting back to more practical matters.

"I don't know, it's crazy, but I was just thinking about pizza. Is that silly? Pizza in Egypt?"

"If they have *Starbucks'* coffee and cheeseburgers, why not pizza? It couldn't hurt to ask."

I catch the server's attention and ask him if we could order pizza here.

"Yes of course," he says with hardly an accent. "What would you like?"

My odds of capturing Linda's heart have just narrowed. If I'm lucky enough to find pizza in the middle of Cairo, anything's possible!

"So how is your wine?" Linda asks.
"I like it . . ."
"You're not just saying that?"
"No, it's alright; really! It's just fine."

We order two pizza slices each and they eventually arrive, fresh and hot. We are pleasantly surprised and dig in immediately.

"There's nothing like Syrah and pizza!" says Linda. "Let's see . . . red wine with meat, white wine with fish, and Syrah with pizza. Perfect!"

Linda finishes her first pizza slice and unfortunately begins to focus on less frivolous matters. "What do you give Syed's chances of getting the third tablet?" she asks, derailing my perfect evening of talking about nothing in particular and enjoying each other's company.

"We should know in the next day or two," I say. "Would you like another glass?" I'm trying my best to redirect Linda's attention. "We should have some time to look around and see the city tomorrow. We did the pyramid thing. What else do you want to see?"

"They should have some nice museums. If Syed feels better, he might show us around."

I notice Linda has begun to refer to "Dr. Azad" as "Syed." That's another good sign that she is getting comfortable around him.

"I remember Syed pointing out the Egyptian Museum of Antiquities the other day when he took me on a tour of Cairo University. I don't think it's very far from here."

"Syed is the perfect guide," says Linda. "His knowledge of the pyramids is phenomenal."

"He seems to be very well-acknowledged in his field. He's even contributed to some of the artifacts in the museum. It should be pretty interesting."

"That second glass sounds good about now," says Linda. "I need to wash out the pizza taste. Oh, I just remembered—if we want a nightcap when we get back, there's only beer. But it shouldn't matter. You know the old saying: 'Beer after wine, you'll feel fine.'"

"You're a veritable font of knowledge!"

We order our second glass of Syrah and I can tell Linda is already feeling warm and fuzzy.

"I thought Muslims didn't drink alcohol," says Linda, referring to Syed's occasional indulgence in a bottle of lager.

"My take on it is a little nuanced," I say. "I've never seen Syed drink anything stronger than lager, and that sparingly. I think he rationalizes his occasional beer because the Egyptians have, since time immemorial, done the same. He doesn't get intoxicated and the ingredients aren't exactly inconsistent with good health. You know that about wine—as long as it's not overdone, it can contribute to a long and healthy lifestyle."

"Sounds like a commercial," says Linda.

"I'm not saying Syed is a saint. I'm just saying hardly anyone practices their faith or philosophy perfectly."

Linda then sums up my argument: "Sort of like the vegetarian-vegan divide."

"Something like that."

Our drinks arrive as our conversation veers off on a metaphysical journey, sip . . . by . . . sip.

"Well, according to the standard party line of Christendom," says Linda, "Jesus was supposed to be perfect. Isn't he an exception?"

"Well, as you said, 'according to the standard party line.' "

"You never said if you believed in Jesus," says Linda, getting serious. "Being a scientist, I just assumed you don't believe in God. I never asked you."

After an uncomfortable pause, Linda pursues the point: "Well, do you?"

"This is where I can't help but get a little opaque," I say. "Usually when I attempt to explain my position several things happen: People think they understand me to mean I categorically do not believe in God. Or they think I have a different version of God than they do but otherwise we're simpatico. Or, finally, they think I am a closet atheist. Unfortunately, none of these conclusions would be true."

"How can that be?" She asks. "Either you believe in God or you don't. It sounds to me like you're trying to hedge your bets."

"Not at all. I've simply placed more than one bet at the roulette wheel, not knowing which one might pan out."

"Isn't that the very definition of hedging?"

"Maybe."

"Okay, I'll cut you some slack," says Linda. Then, with her now trademark gesture—eyebrows up, right hand twirled—she says: "Let's have it."

"I don't suspect there is a singular, self-contained deity calling the shots. I could be wrong but I don't think so—current evidence does not support such a view. Of one thing, however, I am irrevocably certain: intelligence is manifest in the natural world; in the origin and evolution of the universe; and the complex functioning of life processes. This is, in my opinion, irrefutable. And this intelligence—an expression of forces not currently fully understood or embedded into the very fabric of the universe itself—is either directly guiding the evolution of the universe and all it entails or, by default, generating, by its intrinsic nature, an executable program inhered in its very structure and composition."

"That's easy for you to say," says Linda.

We both laugh.

Sip . . . sip . . . sip.

"So, what about religion then? Under your formulation, I don't see how it could be of any use to you."

"On the contrary. There is much in religion, in the sacred writings, that is not only helpful but which I fully incorporate into my own life."

"You mean their teachings? But how could they help you if you don't even believe in the tenets of the religion?"

"But I didn't say I didn't believe in the philosophy or the underlying principles of religious beliefs. There is a great deal in the precepts of religion that can benefit most anyone who meditates on them and incorporates them into his or her life."

"Like the Golden Rule?" she offers.

"Yes, among many others. But it isn't just specific teachings that are important. Sometimes the way a person lives his entire life can serve as an inspiration. Gandhi, Buddha, and the Dalai Lama come instantly to mind. Or Indira Gandhi, Susan B. Antony, and Joan of Arc if you prefer."

"I can't argue with that," she says. "It makes sense, I suppose.

"What about you?" I ask. "What gets you through the night?"

"My parents used to take me to church when I was little but it never took. I just thought it was all a bunch of hocus-pocus. I remember watching televangelists on TV and thinking, even when I was just nine-or-ten years old, 'what a crock!' "

"So, what philosophy do you wrap yourself around?"

"I just live from the inside out."

"I've heard that expression before, on the radio or somewhere. What does it mean?"

"It's a way of living your life that concentrates on what really matters. The basic idea is that events over which we have no control can change in an instant, so, armed with such realization, we need to live more purposefully; to integrate our inner and outer lives—live from the inside out. It's a way of finding peace within. There's more to it but that's it in a nutshell."

"Are you sure you're not a throwback hippie? C'mon, fess up! You're a closet flower child. Admit it!"

Linda laughs, followed by my laughing; followed by both of us cracking up.

"I guess we'll both find out who's right when we reach the pearly gates," I say. "Or not."

We finish our drinks and call Emile to pick us up.

It is late when we arrive back at Syed's house. Emile gets out, goes up to the front of the house, and punches in the security code to let us in. I guess Syed trusts him. No matter; we are both happy to be back.

Linda and I go inside and Emile says goodnight and drives off.

"Where does Dr. Azad keep his stash of lager?" Linda asks, still in tipsy mode.

"Are you sure you want a beer after—"

"Just one," she says. "A nightcap. And then off to bed . . . pinky swear."

My refusing Linda would be like a kid refusing candy. "Okay, but just one and then we should hit the hay."

"I'll pulverize that hay!" says Linda.

Against my better judgment, and far from the first time, I cave in and get us a beer.

We sit on the sofa near the large window.

Linda takes a drink. "So, you and the doctor sleep upstairs?"

"Dr. Syed's room is upstairs and overlooks the courtyard in the back and mine is down the hall just above your room. Both of our balconies face the East and—"

"So, you're on top of me?" says Linda, the innuendo screaming loud and clear.

"In a manner of speaking," I say, feeling a little uncomfortable. Where is this going? Linda is in a playful mood, the signals are crystal clear; and yet, I can't shake the feeling that something just isn't right.

Linda takes another drink. "It's late," she says. "Dr. Azad must be asleep."

"I think that's a good idea," I say. "I should put these beers away. I don't want to leave a mess in the living room." I try hinting that we should call it an evening. Linda had hardly eaten anything and the wine and beer are taking their toll.

Calling out her secret weapon, her dark blues, she looks at me and asks, in her softest voice, "Would you take me to my room, Jake?"

It doesn't take Sherlock Holmes to figure out where this is going.

We walk down the hall to her room and she opens the door. I stand there, motionless, waiting for the next shoe to drop.

"Don't you want to come in?" she asks.

My feelings for Linda are overpowering, competing with that familiar pang of conscience I always get when I'm about to go against my better instincts.

"It's late," I say. "And it is Dr. Azad's home. I don't think we should—"

Linda kisses me gently on the lips and I am just about to lose it when she says: "Okay Jake, I'll see you in the morning. Good Night."

I feel relieved, sexually frustrated, and confused, but not necessarily in that order.

Walking back toward the stairs to go up to my room a thought hits me: A man's libido is very much like a weather forecast predicting a fifty percent chance of rain. When it comes to a man's sex-drive there is a fifty percent chance he will rein it in and do the right thing. Given the same circumstances on any other night, I might have done the opposite. Our social conscience can be so fleeting, ephemeral—Dr. Jekyll or Mr. Hyde; gentleman or rogue; friend or foe.

Chapter Eleven: Convergence

U.S. Egyptian relations remain tenuous. Overtures toward Russia by the Egyptian President hinting at a strategic partnership have caused tensions in Washington. Continuing trade settlements between Egypt and Russia in Egyptian pounds and Russian rubles, respectively—bypassing the dollar—adds further complications to an increasingly tense situation. Add in the fact that the presidents of Egypt and Russia have a strong bond of friendship and mutual respect, and you have the ingredients for an international brew of intrigue and heightened hostility that does not bode well for America.

I wake up early, Syed earlier, and Linda late.

When Linda joins Syed and me in the living room she looks cheerful, rested.

"Good morning gentlemen," she says Linda. "What's on the agenda for today?"

Either Linda doesn't remember our brief romantic interlude from last night or is putting on a happy face to put it behind her. In any event, she wouldn't say anything in front of Syed. I guess we'll revisit last night another time.

Looking at me, Dr. Azad suggests we make a day of it and tour the city.

"I'd love to see the Egyptian Museum of Antiquities," says Linda. "It must be a storehouse of beautiful treasures and artifacts."

"Indeed it is," says Syed. "It has a very long history, going back to the nineteenth century. It suffered some damage during the so-called Revolution but it weathered the storm rather nicely, considering. Shall we put that on our schedule for today?"

"If you're feeling up to it," says Linda. "We could make it another time—"

"I will not hear of it. I feel fine. It should be very enjoyable and I could use the exercise. I have been neglecting myself lately."

Syed excuses himself and goes upstairs for a moment, giving me a chance to find out what, exactly, Linda remembers from last night.

"Did you go to my room last night?" asks Linda.

"Yes," I say, wondering if that is all she remembers. "You had a bit to drink and not very much to eat so I wanted to make sure you made it to your room alright." I'm trying to play down any romantic connotation she might have remembered and, at the same time, minimize her responsibility for getting—let's just say—tipsy.

"Did we kiss?"

"Yes, it was just a little peck, nothing more. The way a loving sister might kiss her brother."

"Not in my family!"

I dare not add a word for fear of hinting at what happened last night. If another opportunity comes my way, and Linda is stone-cold sober—and if she still feels like inviting me to her room—the proverbial wild horses couldn't keep me away!

Syed comes downstairs. "You know," he says, "the museum is quite extensive and we will have to be very strategic as we cannot possibly do it justice in a single visit."

"We trust your judgment," I say. "You lead and we will follow."

Syed offers us something to eat for breakfast but Linda and I defer. "All I need is my coffee and I'm good to go," I say. "What about you Linda?"

"A late brunch would be fine," she says. "I'm with Jake. Just pour some coffee down my throat and I'll be fine."

"I have prepared an American brew. I think you will like it. Emile picked it up for me from your American café. The one you Americans are so fond of."

"Starbucks?" Linda asks, hopeful. I can almost see Linda salivating at the prospect of having good old-fashioned American coffee in the morning. I am just as eager as Linda to have a cup of familiar brew. I appreciate Turkish coffee, if and only if, I have no other options. We are creatures of habit. I imagine Syed would feel the same revulsion about "American" coffee.

"Yes," says Syed. "It is on the front of the package. I do not know if you will like the blend—"

"Doesn't matter; whatever you have," I say. I do my best to sound indifferent and not give away my sheer delight in ditching the Turkish version of coffee for the *real* thing.

When Syed looks away for a moment, Linda catches my eye and holds up two clenched fists while exposing her teeth in a wide, fake, grin. It is no surprise she is as happy as I am to forego the Egyptian "house blend" for a more suitable roast.

Linda and I savor our coffee. It feels like—for just a moment—I am back at home on the patio having my morning coffee. Smell is the most powerful sense in evoking memories. I quickly snap back to the present, the memory vanishing as quickly as it had arrived.

Emile drives up and we head off for the museum.

It takes us less than ten minutes from Syed's house to reach the museum. I am beginning to suspect that everything you need in Cairo is within thirty minutes or so from Syed's home.

It's almost nine o'clock and the museum is just about to open. We get out and wait a few minutes, taking in the breadth and scope of the building, appreciating its neoclassical style and noting the several flags on display on the rooftop and the palm trees framing the front entrance.

We enter through the gates and walk over to the check-in area just to the left. After the check-in procedures are completed, we proceed to the ticket kiosk. I note the wide disparity in prices for locals versus tourists but have no complaints. It goes with the territory, I guess.

Once inside I am immediately struck by the huge, almost cavernous expanse of the museum. There are hallways too numerous to count going everywhere and nowhere as far as I can see. It is a good thing we are with Syed; otherwise, we might easily wander around aimlessly for hours. There seems to be a method to the madness; tour guides available to happily take you on a prescribed route, chronologically, from collection to collection. Syed is having none of it. He takes matters in hand and takes us to the more interesting exhibits. It doesn't seem to matter that we are untethered from any group; Syed strolls about the museum as if he owns it. Maybe he does for all I know. None of the security or other museum personnel approach us. It is, to use Linda's phrase: "Cool!"

After showing us countless exhibits, explaining each one in great detail, and demonstrating a level of knowledge and expertise that any run-of-the-mill guide could only hope to achieve, he takes us upstairs to show us the pièce de résistance: The treasures of Tutankhamun.

Syed tells us about the extensive itinerary the exhibit had undergone, traveling around the world, making stops in the U.S., Berlin, Japan, France, and Russia as well as our neighbors to the north in Canada, among others. Neither Linda nor I had seen the exhibit and were anxious to hear the details, especially from such an eminent Egyptologist as Dr. Syed. By international convention, the exhibit is the property of Egypt and any other country privileged to host the display would do so on loan from the Egyptian authorities.

Syed briefly mentions the Valley of the Kings, explaining to us that it is a valley in Egypt where for nearly half a millennium—from about the sixteenth to eleventh centuries BC—Pharaohs and nobles authorized the building of tombs. He says the mummy of Tutankhamun is the only one from the valley that still rests in its original sepulcher.

He then mentions its discovery by Howard Carter and how Carter immediately recognized the scope of the find. He explains the excavation process which is fascinating in its own right. He says it took ten years to complete the excavation of the tomb from the time of its original discovery.

This is all very interesting to Linda and me, but let's faces it, what we want to hear about is the treasures of the tomb. And Syed does not disappoint. After mentioning the known details of the untimely death of King Tut, he commences to give us a rundown of the artifacts and treasures "sent" along with the King on his journey to the afterlife. Syed explains there were over a hundred different articles "attached" to the King's body. Their purpose: to provide the King with protection to safely carry him to the "netherworld."

From the golden mask to the burnished hands of gold, to the spell inscriptions—created and carefully positioned according to the directives of the Book of the Dead—Syed's insights and inside knowledge is both exhausting and, at the same time, absolutely fascinating.

On our way out of the museum, we hear what sounds like an explosion. "What the hell was that?" I say. "Are we under attack?"

Syed remains calm while Linda and I are near panic. How does he do that? Remaining so calm in the face of what seems to me imminent danger?

An announcement blares out over the PA system informing us the museum will be closing.

"Wouldn't we be safer in the museum than out on the street?" I ask, incredulous that our safety is deemed more secure outside where, if there was an explosion, we would be thrown into the thick of the commotion.

"Not to worry," says Syed, in the same tone he might have used when ordering Turkish coffee. "It is normal procedure."

We leave the museum as directed and Syed calls Emile. We look around but don't see any clues as to what may have caused the disturbance.

"It is most likely some local radical group causing trouble." Syed's reassurance goes a long way in calming my nerves. Linda seemed less on edge as well and we soon chalk it up to a local disturbance as Syed had suggested and decide to enjoy the rest of the day.

We go to a nearby café and have a light meal. Neither Linda nor I bring up the subject of locating the third tablet. If Syed has any news to report, he will let us know soon enough.

Syed's phone rings. This time he doesn't excuse himself but takes the call while sitting with us. I can't make out much of what is being said. Syed is speaking his native tongue. I'm certain however that I heard the words ruble and pound several times. I don't think I'm mistaken.

Syed hangs up. He has that look of someone who is thoroughly satisfied with himself. I know because I felt that way once many years ago when I first learned I had made tenure. I went around campus telling everyone I didn't need my brain anymore. Most anyone in academia is familiar with the joke: The juvenile sea squirt wanders through the sea searching for a suitable rock or hunk of coral to cling to and make its home for life. For this task, it has a rudimentary nervous system. When it finds its spot and takes root, it doesn't need its brain anymore, so it eats it! It's rather like getting tenure.

"Good news?" I probe, budding in where I don't belong . . . again.

"Yes, very good news."

"Did I hear you mention rubles and pounds?" I ask, risking insult for my impertinence.

"Yes, I believe I have a very good play on the ruble."

"Should I jump in?"

"Only if you are prepared to be a 'player.'"

I get the feeling Syed is just being polite, knowing I couldn't *play* in his league. He is probably right, but I'm still interested. "How big of a player?" I ask.

"Well," he says, taking the question seriously, "when you invest in foreign trading you must be prepared to risk quite a bit; otherwise, there is very little to gain. Of course, there is always much to lose as well. Either way, it is dangerous waters and many sharks are waiting to attack if you do not know what you are doing."

Syed's use of metaphor is impressive. Pressing further I ask: "So what is a good bet? What do you think is happening with the ruble that could make you a profit?"

"If you seriously want to know, I will tell you."

"Please."

"Okay . . . now first of all you must be very serious about your ability to withstand a substantial loss if you are on the wrong side. Fortunes come and they go very quickly in this market, you must exercise great caution. I do not usually invest in this foreign exchange business; it is very tricky. However, I believe some events may take place soon that will affect the ruble and give me a slight advantage. As you must know, timing is everything—when to get in and when to get out. The market can sometimes be very . . . finicky."

"Which side of the market in rubles are you on?" I ask, assuming he must be shorting the ruble.

"I am on the positive side," he says. "I have a position that should develop quite nicely over the next several months or so."

"I would have thought you would be shorting the ruble. Isn't betting on Russia a gamble?"

"Yes, of course," he says, "that is just what I am hoping many investors will believe. They will be playing right into my hands. When the market goes south you should go north. You claim your position when others are certain they know what they are doing; they usually do not. There is nothing more rewarding than to take advantage of other people's certainty. You must read up on this Nassim Taleb gentleman; he is very clever."

And then, after a moment of reflection, he says "Of course, I could—how do you say 'lose my shirt'!" He then laughs, adding: "But I have learned not to be too greedy. I have a good position but, if it does not produce results, I am prepared to take the loss. What is it that you say, Jake? 'No risk, no reward.' "

Images of the Bilderberg Group spring to mind. Syed is obviously a member of an international cartel manipulating the market, taking pity on an inconsequential physics professor and tossing me a crumb. Here I go again! And I was doing so well.

After we arrive back at Syed's home, he asks Emile to join us inside and makes a couple of calls. "The disturbance near the museum was, as I suspected, some local radicals seeking attention. There was some minor damage but it is nothing serious."

Syed's casual delivery of the news speaks of his growing acclimation to modern life in the Middle East: Pyramids, sphinxes, exotic nightlife, desert storms, the Arab Spring, terrorism—c'est la vie!

In another life, I could see Syed working for the CIA in counterespionage or counterintelligence—nerves of steel; an implacable character; and an indomitable spirit. With Emile's military experience, Syed's firm resolve and my . . . Oh never mind.

The call that could tie it all together finally comes in. Khaled had spoken to Syed and informed him where we could collect the remaining tablet. It is now late afternoon and Khaled is seeking a meeting with Syed as soon as possible.

Syed looks at me tentatively and takes a deep breath. "This Khaled fellow says he wants us to meet him at some address in the city, some apartment. I wrote down the address. It is not in so good a neighborhood I am afraid."

"Can't we just meet him at the club like we did last time?" I ask.

"If the decision were up to Khaled it would be no problem. Unfortunately, the tablet is in the possession of his former accomplice. I do not think he trusts Khaled with the merchandise. Khaled says his former partner is wanted by the authorities. I spoke to him about a possible meeting that would be more open but he would not agree. I suspect the meeting place is a temporary rental that is not connected to them in any way. I got the impression they need to move quickly. Perhaps their sense of urgency will give us the advantage."

I know this is going to be another one of Syed's one-word answers but I have to ask: "Do you think it might be dangerous?"

"Yes."

Maybe I could stay here with Linda while Syed and Emile retrieve the tablet. We would just be in the way. What's the matter with me? I can't let Syed and Emile take all the risks. What would Linda think of me? What would *I* think of me? Well, if we're going, better to do so in the daytime before it gets dark.

"We should go pretty soon before it gets dark, don't you think?"

Syed is clearly of like mind. "Yes, we should go." Syed then looks briefly at Linda and then at me. "It is not a place for an American woman. I hope you are not offended."

Linda knows when to assert her independence and when to bow to custom. "No, that's alright. I know you are thinking about my safety. I'll be just fine."

"Jake, perhaps it would be best if you—"

In the most assertive voice I can muster, I say: "I'm going with you, Syed."

I look over at Linda and then back at Syed. Linda doesn't move or say a word and Syed is already turning to leave. "We'll be back very soon," I say. And off we go.

It is obvious Emile is now "in the loop." It's okay with me; we might need his special brand of *expertise*.

After driving for some time, we reach the apartment building where we had agreed to meet Khaled and his "partner." We get out and walk up to the entrance. Syed and I exchange glances. Emile stands with a ramrod posture, gaze fixed on the doorway entrance, waiting for a nod from Syed. We go in.

When we reach the apartment, Emile knocks on the door and it opens immediately. Neither Khaled nor his partner dares offer a greeting other than to sheepishly invite us in.

My tension is quickly relieved. Syed seems to be in command of the situation and Emile's presence gives me further assurance that things will not get out of hand. They don't. We complete the transaction without incident. Syed's intuition had proven correct—Khaled and his former accomplice are in a hurry and we had the upper hand.

Syed tucks the tablet under his right arm and we head back toward the car. Syed gets in and sits down.

Emile and I are standing by the car door when two men come up to us. One of the men attacks Emile. Emile, anticipating a confrontation, dodges the first blow but then takes a punch to the stomach. As the other man turns toward his partner, I grab him from behind and hold on as tight as I can.

Syed is by now out of the car and sees me struggling with the assailant. Syed cocks his right shoulder back, raises his right arm making a tight fist, and slugs the guy so hard he knocks both of us off balance. The guy is out cold, his body going completely limp as I lose my grip under his weight and we both fall to the ground. I glance over at Emile. He has recovered quickly and fells his opponent. Emile's assailant lays on the ground for a moment and then springs to his feet and runs off.

I am shaken. I can't believe what happened. One moment we are about to get into the car and the next we are fighting for our lives.

We all look at each other, processing what had happened, palpably relieved that we are still standing, relatively unscathed.

Syed and I sit down in the back seat and Emile drives off.

I am still shaking so badly I stiffen my arms and place my hands on my knees to steady myself.

"You did very well Jake."

Syed's statement is just what I needed to hear. I thought I had acted cowardly; all I could manage to do was grab the guy from behind and try to restrain him. I am surprised I had the gumption to even do that. At least I hadn't run away.

Syed looks at me the way my father used to when he knew I needed reassuring. "They thought they were dealing with weaklings. I guess we showed them, eh Jake? I told you we could count on Emile's military training."

"Yes," I say, "he was terrific. We were lucky to have him with us."

Syed then says what I had been thinking all along: "It is a good thing they were not carrying weapons. It might have turned out very badly."

We live our lives on a razor's edge. One moment we are congratulating ourselves for some perceived achievement, living a life we had previously only dreamt possible, the next our entire world is shattered—our lives, our hopes, our dreams—tossed into the garbage like so much refuse.

On our way back Syed suggests it might be better not to mention our "incident" to Linda. Even though I am anxious to share my tale of bravado with her, I know Syed is right. No sense upsetting her. I can always relate the tale after we have completed our work here and returned to the states. By then I will have re-remembered the event enough times that the details should be substantially improved.

We get back to Syed's home just before dark. Syed sends Emile away and we go into the house. I don't see Linda so I assume she is in her room, perhaps taking a nap. Syed and I relax with a cup of tea and recap the day's events.

"So, Dr. Jake, are we now 'out of the woods'?"

"I hope so. I don't think I could go through that again. By the way, that's quite a wallop you carry! That guy is probably still napping."

We both have a good laugh. Just the therapy I need.

Linda comes in to join us, putting the kibosh on our conversation. It's okay with me; I prefer a new venue—anything to divert my attention away from that traumatic event.

"How did it go?" Linda asks, impatiently. "Did you get the tablet? Were there any problems?"

"Everything went okay," I say. "It was a rough neighborhood but we didn't have any problems," I lie.

"And the tablet?"

"It's secure; we got it," I say.

Linda looks at me like a little girl wondering what I have done with her Christmas present.

"Well where is it?" she asks.

"Syed put it away in a safe place," I say.

Linda is having none of it.

"Well let me see it! I didn't come all this way for *nothing*!"

I look over at Syed.

"She is correct," says Syed. "She has come a very long way. She should see the tablet. I will bring it."

Attempting to bring things back to a more even keel, I ask Linda if she had been able to get some rest while we were away.

"I tried to lie down for a few minutes but couldn't sleep. I'll catch up soon enough. I'm sorry I came on a bit strong Jake. I was worried about what might have happened."

Of course she was. I should have realized Linda is just on edge . . . waiting . . . not knowing what might happen.

"It's okay," I say. "I understand. It was pretty intense for a while but everything worked out alright. I'm glad it's all over. Now we'll be able to focus on where we go from here."

Syed returns to the living room with the tablet and places it on the large oval table in front of us.

"Look at the color!" says Linda. "This one is definitely different from the others. I wonder what coloring process they used. And look—the image on the plate is shifted or jiggled. What could that mean?"

"It is difficult to say," offers Syed. "We will need to examine it further. It is different from the other tablets in this respect." There is a heaviness in Syed's voice. His face looks drawn, worn even. Our brawl with those thugs had taken its toll.

"We'll have to figure out our next step," I say. "It looks like we have all of the information we're going to get from Khaled and his cronies. How do we go about getting the tablets further tested to determine if the gaps might reveal anything?"

"I don't see any other way than to somehow get them back to the states," says Linda. "Unless there is some other technology we don't know about that is available locally. What do you think Dr. Azad?"

"I do not believe the technician at the university knows any more than what he told me. He offered no ideas when I asked him about the divisions in the tablet. But the process may not be as difficult as we believe. Perhaps we should look into the matter further before we decide what to do."

"There would be some risk if we do need to take them back to the U.S. for further examination," I say.

"Not necessarily," says Syed. "There are ways to provide security. We would need to be very careful but it should be possible."

I can't see the point in pursuing the matter any further. Until we have something more concrete to go on, further discussion won't get us anywhere. "I wanted to thank you for the tour of the museum, Syed. I really enjoyed it; especially the King Tut exhibit."

"Yes," says Linda, "it was wonderful."

"It was my pleasure. I must apologize, however, for the abrupt manner in which our tour ended. Such drama is not usually included."

We laugh. I am glad to see Syed regaining his sense of humor.

Linda, recalling my previous tip, brings up one of Syed's favorite subjects. "Jake told me you have a set of classic American movies."

Syed's enthusiasm picks up. "I have a small collection," he says.

"Would you mind if we watched a movie?" says Linda. "It would be nice to have a little distraction; take our mind off of things for a while."

"I have this very good movie," says Syed. "I am sure you will like it. It is called *Casablanca* and . . ."

Chapter Twelve: Nooks and Crannies

All the pieces are coming together. We have three tablets—two representations similar, one somewhat different: the Schrödinger equation, a double-stranded polymer molecule, and the cosmic background radiation. What do they have in common, if anything? Do they represent more than meets the eye? Is there some hidden code or information embedded within or underneath each image? Is the "shifted" or "jiggled" image on the tablet containing the cosmic background radiation relevant? And most important, do we have the requisite tools—forensic, intellectual, and analytical—to solve this mystery? As these questions churn in my mind, I become ever more determined to decipher the deep enigma confronting us. If there are answers I will find them.

I am beginning to adapt to the rhythm of Cairo. In many respects it is no different from any other large city—people getting up in the morning and going to work amidst modern skyscrapers and the constant blare of traffic; negotiating the highways and subway system. These modern-day aspects, in part a reflection of Western influence, meld together with Islamic mosques, pyramids, and the occasional processions of camels and donkeys along major thoroughfares.

    The morning light streaming into my room announces the start of another day in my search for answers.

It is clear to me: We have to contact Boston University and speak to someone in the research department working on the scanning technology we need. This is Linda's project.

I shower and put on a fresh set of clothes and go downstairs. As usual, Syed has already begun his day. He is sitting in the living room reading an Arabic newspaper.

"They still print newspapers?" I ask, hoping to start the day on a friendly note.

"It is always something. These people will never join civilized society."

Syed has a look of disgust on his face. I better change the subject.

"What do you enjoy doing in Cairo when you are not entertaining guests at home?"

Syed's disposition changes instantly. "I sometimes go to the theatre. There is a show I am looking forward to seeing."

"What type of show is it?"

"It has some dancing and singing. I go more to relax; to change the routine."

"What's the name of the act? I don't suppose I've heard of them."

"They are with the Heritage Dance Troupe. They are performing in Cairo presently. They were here some years ago but I missed their performance."

"For some reason, I don't picture you liking musicals."

"Occasionally I do. It depends on the performance. They are very good. Perhaps, if we can find some time before you leave, we will be able to take in a show."

"I'd like that," I say, relieved that we have managed to avoid any talk of politics.

Linda joins us just as the doorbell rings.

Syed lets the housekeeper, Abasiama, in and speaks to her for a moment. She goes directly upstairs. Syed apparently does not feel the need for introductions.

"Did you sleep well?" I ask Linda.

"Much better. I think I'm starting to get in sync with the time difference."

"Linda," I say, anxious to move our quest along, "I think we should go ahead and follow up on the new scanner. What do you think?"

"Sure, I'll make a call a little later. They should be available in about—"

"Let's see, it's seven here. Nine hours minus three would be six, subtract . . . It should be about one in the morning east coast time," I say.

"Jake, the human calculator!" she says.

"That should give us some time to map out our strategy," I say. "We need to plan exactly how we're going to handle this discovery."

Syed nods in agreement. "You are right. We should be careful about how we proceed."

"If the research department at Boston University gives us the go-ahead," says Linda, "we'll need to figure out how we're going to get the tablets back to the states. They will have to be secured somehow. And then there's the matter of confidentiality. As I mentioned before we may have to include one of the research members in our little secret. Of course, if we can't get permission to use the new technology, all bets are off. We'll just have to work with what we know."

"Syed, is it possible that a newer scanning technology might be available here, outside the university? Perhaps there are some businesses or research facilities somewhere." I'm grasping for straws but it's worth a shot before we transplant our operation back to the states. I also want to keep Syed engaged and working with us directly.

"That is worth looking into," says Syed. "I will see what I can do. You mentioned Boston University. That is interesting."

Linda looked puzzled. "How do you mean?"

"There are many facilities throughout the world that are working with us to uncover the secrets of ancient Egyptian civilization. There are such facilities even in your country. A research institute is not so far from Boston University. Our department has worked with them on occasion. It is even possible that the university in Boston has been in touch with them."

"What institute is that? Would they have a branch here?" Linda is reeling in Syed's thoughts to narrow our search, and possibly the geography as well.

"It is called the Ancient Egypt Research institute or something. I do not remember the exact name. I do not know of any offices here in Cairo but I cannot say for sure. They may be working on something new. Perhaps they even know of the work of the university." Then, almost as an afterthought, he says: "There may be other methods we can use. I am not an expert on scanning technology. I am more acquainted with dating methods and establishing chronology. And as far as the gentleman at the university who scanned the tablet, he is nothing more than a lab technician. I doubt he knows any of the details about what he told me. He was unable to stop his machine at different levels on the plate. He may have thought he was looking at an obstruction of some kind, shadows on the image possibly."

"So, there might be some other type of scanning technology we could use without having to rely on Boston University?" I am piggybacking off Linda's idea: Maybe we could use some alternate technology that might serve the same purpose.

"You would be surprised what we can do," Syed continues. "We have examined well-preserved mummies dating back many years. We have even found tattoos on some of them. That was a surprising find."

Linda laughs. "Wow, that's amazing. And you bring up a good point Dr. Azad—"

"You may call me Syed. A friend of Jake's is a friend of mine."

"Syed," says Linda. "We don't know everything about scanning techniques. Maybe there is some device we can use right here in Cairo."

"There are special cameras that can take pictures exposing additional features underneath a surface. The medical field has—what do you call them?—MRIs and other scans . . . No, MRIs would not work."

Linda and I look on as Syed continues his internal dialogue.

"We may only need something that can expose three-dimensional properties so we can—"

"I think we're onto something," I say, impatiently. "How can we follow up to find out?"

"I will make some inquiries."

Syed's assurance that he will look into it further is all I need to hear. He has connections, resources.

"I should take one of the plates for inspection. Do you have any preference?" he says.

"Well," I say, "the first plate is the only one examined so I think—"

"Yes, of course," says Syed.

Syed looks at Linda and me, our impatience easy to read. "Let me see what I can do," he says, excusing himself. "I will be right back."

Syed leaves the room and goes upstairs.

"I guess he doesn't want us to see him pulling strings," says Linda. Her eyes then dart in my direction, her face lights up. "State secrets!"

I laugh. If only she knew.

Abasiama comes downstairs and starts walking toward the kitchen and dining area.

"Hello," I say.

"How are you?" she replies, timidly, and in passable English.

"My name is Jake and this is Linda. Your name is Asiana?" I ask.

"Abasiama," she replies. "It is good to meet you."

She then sees Syed coming down the stairs and takes off toward the kitchen to complete her chores.

Syed comes into the living room looking very pleased. "I have some very good news!" he says. "I have spoken to the American Research Center and they will be able to assist us."

I am curious to ask Syed how he managed to make arrangements so quickly, and so early in the morning, but why bother. Besides, if I know how he was able to accomplish such feats, he would simply have to have me killed. Now Jake, stop it!

Syed explains that the American Research Center is in Garden City and they would be glad to accommodate his request.

"Don't tell me," I say. "Garden City is less than fifteen minutes from here."

"It is nearby, yes," says Syed, missing my attempt at humor.

"How does the process work?" asks Linda, curious if there is some new technology we are unaware of.

"As it was explained to me, the scan can—what words did he use?—dissect . . . or remove the outer layer and see what lies beneath, step by step. They can see within millionths of a meter. It should be all the resolution we need."

"This *is* good news," says Linda. "We might finally be able to get to the bottom of this. Unless"—Linda pauses for a moment—"there isn't anything else."

"Well, at least we'll know," I say. "Either way, we still have a lot of work ahead of us."

"I have sent for Emile and he should be here shortly."

Emile seems to be at Syed's "beck and call." Perhaps there is some deeper connection between them other than chauffeur and boss. Maybe . . .

Linda, thankfully, interrupts my musings. "I wonder what we'll find," she says, unwilling to let it go.

"Remember, 'All good things come to those who wait,' " I say, expecting Syed to chime in to deconstruct my latest contribution.

He doesn't take the bait.

Syed "offers" us a sampling of Danish pastry; after which, he takes the tablet and goes off with Emile to the American Research Center.

"There *has* to be something more to uncover," says Linda. "We can't possibly have all the clues."

"Well, as I see it, these tablets already reflect some type of superior technology or foreknowledge. It shouldn't be surprising if they—whoever 'they' are—can embed further information in the tablets."

Abasiama comes into the living room and it is apparent our presence is keeping her from completing her chores.

"Why don't we sit outside for a while," I say, nodding toward Abasiama.

"It looks pretty nice out," says Linda, taking the hint.

We sit facing the courtyard, enjoying our coffee and each other's company.

"Where did Syed say the first tablet was found?" asks Linda. "I forget if you told me."

"Somewhere near Giza," I say. "But that came from Khaled and I don't think we can put much stock in anything he says."

"I don't know if it matters. It might."

"So, what do you think of Cairo, so far?" I say, doing my best to divert the conversation. We won't be able to resolve anything until we have more information anyway. We will have to wait for word from Syed.

"It's different," says Linda.

"Now there's a world-shattering insight!" I say, with my characteristic sarcasm. Linda knows me well enough by now not to take offense.

"No, I like Cairo. I'm sure there's more to it than the usual tourist traps."

"You may not want to see the underbelly; it might surprise you, and not in a good way."

"Oh, I know there's a darker side," says Linda. "Any big city has its murky elements."

"But you're right," I say, bringing us back into sync on our take of Cairo. "There's quite a bit we haven't seen that we might like. The Cairo Tower looks pretty cool. I looked it up and it says they have a restaurant near the top that spins around every hour or so."

"Like the Space Needle in Seattle. Have you ever been there?"

"Yeah, I went there several years ago. Nice view of the ocean and surroundings. Expensive as I remember. The servers probably make more than I do."

"I underestimated how hilly it is," says Linda. "I parked near the ocean thinking I would take a stroll to get the hang of the city. It nearly killed me. By the time I made it to the Space Needle I was exhausted."

"Take your choice—the seven hills of Rome or Seattle. Did you go on the boat ride around the harbor?"

"No, I skipped that. I should have gone."

"Are you still on the tenure track?" I say, attempting to keep our conversation away from any discussion of "the discovery."

"As long as I don't have a bad year, I think I'll be alright. I know publishing isn't the 'be-all and end-all' of academia, but I still try to keep my production at an even keel, trying not to underperform from one year to the next. If I could just get my name on something, some idea or new concept, I think I would be home free. I need to concentrate on one core idea. I have so many competing interests, so many 'irons in the fire,' I'm having trouble pinning down a single idea."

"What idea most interests you?" I ask, trying to be helpful.

"There's a lot that has already been written about the bioethics of genetic research, but we still haven't reached consensus on basic guidelines. Somatic and germline genetic research is advancing so quickly we haven't even mapped out a clear strategy for dealing with the ethical and moral issues involved. Way back in the nineties we were already introducing gene coding for resistance to neomycin—"

Linda gives me that, "You don't know what I'm talking about" look.

Trying another approach, she continues: "All I am saying is that we do what we always do: rush headlong into new fields of research to see where they will lead and, when we find something interesting—something that might work—we set off on an uncharted course with little or no thought to what we're doing. Somebody's got to take this subject seriously . . . No, that's not fair; there have been some honest and sincere efforts to deal with the problem. It's just frustrating, that's all."

"One of my old mentors used to say, 'If it has to be, it's up to me.' " There I go again offering platitudes when Linda is being serious and needs my best advice.

"Linda, in all seriousness, I can't think of anyone more qualified to tackle this problem than you are. You're smart, dedicated, and you've more than earned a right to have a say. You just need to concentrate, the way you do on any problem, and you'll find the answers. Why do you think I confided in you and asked you to join in on our search?"

My little pep talk works magic, as I had hoped.

"You're right Jake. I just need to buckle down and concentrate and put all of the pieces together. Thanks, I can always count on you."

"It is *what I do*!"

Linda laughs at my pathetic attempt at humor. I guess it was pretty lame. I had never asked Linda why she is only now pursuing tenure; why she had completed her doctorate relatively late in life. This might be the right time. Linda seems to be in a good mood. So, I ask.

"After I received my undergraduate degree, I took some time off." Says Linda. "One thing led to another and it was several years later before I decided to complete my studies. Right at the time I was ready to return to university, my mother took ill and I needed to take care of her so I put everything on hold. I didn't mind. She took care of me and it was my turn to take care of her. That's how it's supposed to work, isn't it?"

"Well, that's how it used to work. I'm not so sure nowadays."

"Maybe we should go back inside," says Linda. "I think Ab . . . Abas—"

"Abasiama," I say. As if I knew all along how to pronounce her name.

"I think she has finished cleaning the living room."

As we walk back inside to wait for word from Syed, my thoughts turn to Linda and how much my feelings for her have grown since we first met five years ago soon after she joined the staff at Stanford. I hadn't read a book since my flight over. All the excitement over the find is distraction enough for me. Linda's company is a pleasant bonus.

Syed eventually calls. I immediately answer, anxious to hear what he has to report.

Syed speaks in a professorial tone as if he is giving a press conference. "Dr. Banner," he begins, "I have some very exciting news. We have, how do you say, 'hit the mother lode.' There is very much to consider Jake—"

"What is it? What did you find out?" I don't want to hear an announcement; I want the scoop!

"There are many diagrams and mathematical formulations. Some of them are showing very old information and are easy to understand but there are some I do not understand so well. You will need to evaluate them, Jake. My math skills are very good but these will need more work. They are not so clear. The information is arranged in a sequence . . . what you call a—"

"A cascade," I offer.

"Yes, I suppose that is what you would say. They are in line, one behind another, and so on."

Linda is listening intently, trying to pick up any clues as to what Syed is telling me.

"Did the technician suspect anything?" I ask, fearing our whole operation might be blown.

"No, you need not worry. The technician did not suspect anything. He asked me how we were able to conceal a 'recording' in the tablet but did not follow up with any additional questions. As far as the information itself is concerned, I do not think he understands the math. He had no idea of their meaning. I made sure that my visit would be confidential."

Syed is always telling me there is nothing to worry about and I'm always worrying. There must be an equation in here somewhere.

"The documents have" . . . Syed pauses for a moment to collect the right words, "a double image."

"What do you mean?"

"You will see when we review the documents," he says. "I do not think it is a problem."

I dismiss Syed's comment about the image, grateful we can extract more information. "This is unbelievable! Were you able to produce any copies of the images?"

"Yes, yes, we should be able to do so. I will be there soon and we can examine the whole lot. Listen, Jake, this is very interesting. The technician did not discover the data on the tablet. He *read* the information. There is a hidden device in the stone. I do not fully understand the methods he was using. It seemed to me to be an ordinary scanning machine. The important thing is that we now know there is much more data than we have suspected."

I want to ask Syed more questions but figure it can wait. "Okay Syed, we'll see you soon."

No sooner do I hang up than Linda pelts me with questions.

"The technician recovered the information," I say. "Don't ask me how. I don't know. It sounds like there's a lot of data. Syed said he doesn't even understand most of it; not in detail anyway. He knows math as well as anyone teaching at the university, so if he didn't get the gist of what he was looking at right away, it must be fairly advanced. Oh, and he said there are also some older diagrams and computation as well as new ones . . . and the documents have what he calls a double image. Strange."

"So, Syed has the information? We'll have a chance to see it?"

"Yes, he has everything. We just need to sit tight. He'll be here any minute."

"We've done it, Jake. Well, we've *all* done it. It's real!" Linda laughs nervously, succumbing to all of the excitement.

It is at this moment, looking at Linda, I realize two things, without a doubt: I love her . . . and we are officially embarking on the discovery of a lifetime—make that of *all* time!

"I can't believe what's happened," says Linda. "At first I thought maybe we were on to something; it seemed genuine. But now it looks like the real thing. It's difficult to process. What can it all mean?"

"We'll just have to take it one step at a time," I say, the only palliative I can think of on the spur of the moment.

Just then Abasiama is passing through the living room. This might be a good time to learn a little more about her and her country while Syed is still out. I try to never pass up an opportunity to learn about other cultures.

"Abasiama," I call, as she is walking past.

"Yes," she says, timidly.

"Is it alright to call you Abasiama?"

"Yes, it is okay."

"You are from Kenya?"

"Yes. I have been here for some one year."

"Would you mind if I asked you about Kenya?"

Abasiama looks nervously back and forth as if someone might be listening. "It is okay."

"There has been a good deal of violence in Kenya and throughout much of Africa. Is that why you left your country?" Am I being nosey again? Of course, it is *what I do*.

"There is much . . . strife . . . Christians and Muslims continue to fight."

"Do you think there is any hope?" I ask, continuing my interrogation.

"Oh yes, it is possible. Boko Haram is very military and wants to kill all Christians, but there are many Muslims that help us. They have fought for us when the Fulani argue with Christians over the grass and the water for the animals. There are some bad Muslims but not like doctor Azad; he is a very kind man."

Muslims helping Christians; perhaps there is hope after all. The old saying, "You learn something new every day" is certainly true but *only* if you are inquisitive enough . . . and nosey enough, to discover what's new.

Syed arrives, interrupting my debriefing. Abasiama leaves the room and continues her housekeeping duties.

"I have everything," says Syed, in a self-congratulatory tone; well deserved.

"Where are the records?" I ask, eager to dive in and see what discoveries I might find.

"There are many documents. It may take us some time to go through them. Some are not so easy to understand. I have everything on what is called a 'thumb drive.' We should be able to view them on my computer."

At the risk of upsetting Syed, I once again ask him how the lab technician was able to extract the information from the tablet. Could it be as simple as he made it sound?

Syed is unperturbed. "He told me that retrieving the data from the tablet is no different than producing images from the scan, which they do routinely. The scanning machine is connected to the computer and the document images are displayed on the screen. Perhaps there is some property or design in the device, a disc that allows scanning technology to see the images without removing it from the stone."

"I'm not sure what it all means Syed but—"

"Jake, such matters are not so important right now. Let us examine the data."

Syed is right. The technology in the device embedded in the stone was designed to allow easy recovery of the data; even with a scanning device. What is important is the information *in* the disc. We can always examine the disc later to better understand its properties.

Syed directs us to his studio which he had set up as a makeshift office.

Linda and I grab a couple of chairs and join Syed at his computer terminal.

Syed offers me his seat in front of the terminal and hands me the thumb drive.

I insert the drive.

Chapter Thirteen: The Mother Lode

What if we knew everything possible to know: the origin and evolution of the universe, as well as its eventual fate; the complete genetic and neurochemical nature of life; the limits of human knowledge and our inherent ability to achieve further progress toward reaching those limits; the outer boundary of medical and therapeutic applications possible to enhance, augment and sculpt human capacity; and, finally, the overarching *principle* that governs the very nature of life, the cosmos and the limits of human achievement?

"What am I looking at?" I say, trying to get my bearings. "There is so much information. Where do I start?"

"The documents are in order from bottom to top. I can easily tag them if you like so that we will have a numerical sequence." Syed is as eager as Linda and me to glean what we can as quickly as possible from the cache of documents.

Linda and I look on as Syed sequences the documents, the double images clearly visible—mirror outlines clinging to each document; a cascade of empty *shadows*.

"Please," I say, moving out of the way to allow Syed to collate the files.

A few moments later Syed completes the sequencing.

As I review the documents the double images became less distracting and I can finally concentrate on the information. The first document is a straightforward illustration of the Pythagorean Theorem depicting a right triangle, showing a square extending from the hypotenuse and doing likewise with the other two sides, thus illustrating the equivalency of the squared hypotenuse to the remaining, combined, squares of the other sides.

"So, we start with the Pythagorean theorem. Why this particular statement and not another? What do you think?" I ask, my question lobbed into the air for either Syed or Linda to catch.

Syed beats Linda to the punch. "The theorem, or at least the idea, goes back some way; I think at least to the end of the third millennium B.C., during the Middle Kingdom of Egypt. Perhaps it is the oldest calculation deemed to be of any importance. Of course, the Babylonians and Egyptians did not formulate, explicitly, the proof as we are seeing. That was closer to the sixth century B.C., I believe."

"That's very perceptive, Syed. You may be right." To draw Linda into the discussion I ask her what she thinks of the idea.

"That sounds right," says Linda. "Let's look at the next item; maybe we can establish a timeline."

"Excellent," says Syed. "Let's have a look."

We continue to search each file, plowing through the works of Archimedes, Euclid, and many others until we hit upon Schrödinger's equation, item fifty-one—just after item fifty, Brun's theorem, stating that the sum of the reciprocals of the twin primes converges to a finite value.

"How many items do we have in total?" I ask, thinking out loud before answering my own question. "Let's see, there are one hundred and one items in all. The image on the tablet is item number fifty-one. That would be right in the middle. Does that mean anything?"

"Perhaps it is a marker of some kind," Linda suggests.

"Like it was representative of a category," I offer, confirming and agreeing with Linda's conjecture.

Syed concurs and takes Linda's hypothesis a step further. "We have three tablets and each contains different areas of knowledge. The inscription or image on each tablet must, therefore, be a marker, as you said, and I also believe it marks the midway or middle separation in the chronology of events."

"That's brilliant!" I say. "But wait a minute. If the marker for each tablet is the mid-point, then just how far do the documents reach into the future? And why aren't there more . . . or less? Why do they stop where they do?"

We are all now fully engaged, our minds whirring like a fine-tuned automobile revving its engine. If our combined intellects can't solve this quandary, then who? I am reminded of the quote, attributed to many but best remembered when spoken by President John Kennedy, "If not us, who? If not now, when?"

Continuing through the data we discover complete formulations of ideas that were previously, and at best, considered "works in progress."

There isn't enough time to dissect everything but, at this stage, we don't need to. We know what we have: A compilation of discoveries in mathematics and science spanning millennia, from small beginnings to formidable accomplishments in geometry and quantum field theory; from Euclid's elements to cutting edge breakthroughs in cosmology and physics; new ideas and discoveries reaching beyond the present into the future—but how far into the future? That is the question we now consider.

"Most of these other documents contain text," I say. "A few of them just have illustrations but these are self-evident and don't require any explanation. And they're all in English. What do you make of it?"

"My best guess," says Linda, "would be that the information was intentionally gathered for the specific purpose of creating a permanent record for others to see. English is still considered the 'universal language' in the scientific community so that's not so surprising."

Centrist arguments aside, Linda's assessment makes sense and neither Syed nor I feel the need to comment.

The next document catches my attention even more than the others. "I think I found the document that explains how the information was embedded in the tablet."

"That is very interesting," says Syed. "So, we may have an answer to this puzzle. What do you make of it?"

"I've heard of more than one project working on this concept," I say, "but miniaturization and preservation, durability, have always presented insurmountable problems. Historians have been concerned for some time about long-term storage media for documents considered too important to preserve on just any medium. You have to keep transferring information as technology improves and there is always the possibility of losing the data altogether."

I continue analyzing the document as Syed and Linda look on, expectantly. "Give me a minute," I say. "I think I have the gist . . . yes; it makes perfect sense. Storing data involves a durable data-substrate. This is incredible! They've taken a tungsten-coated optical disc and modified it so it can be inserted and concealed in almost any material. Silicon usually fractures fairly easily but they've come up with a coating that protects the disc. The anomalies, or 'gaps,' the lab technician saw when he scanned the first tablet weren't gaps at all. They were simply disturbances created in the limestone due to the scanning device. The scanning machine at the Cairo University lab must have been too old to achieve the proper resolution. The newer scanning machine at the American Research Center made the difference."

"There is a disc in each plate," I continue. "They've taken the scale down to the microscopic level, and I'm not just talking about the information—the disc itself is microscopic. Its heat-resistant properties and size make it practically indestructible. The data is retrieved by simply *scanning* the disc . . . amazing."

Linda finishes my thought for me: "So the information isn't entwined in the stone itself. The stone is more like a casing than a storage medium. There's a disc within the tablet, as we suspected. No matter how sophisticated technology gets in the future, it couldn't possibly use the molecular structure of stone for data storage and retrieval. We're looking at using DNA to improve computer processing and computation capacity, and its ability to retain information is remarkable, but there had to be a better explanation."

"You're not alone," I say. "I couldn't figure out how it could have been possible either. Now we have an answer."

Syed asks the next logical question: "Would it be possible to remove the disc without damaging it?"

"It may be," says Linda. "But we should be very careful at this stage. It might be better to complete our analysis before we attempt to examine it further."

"Yes of course," says Syed. "We should finish the assessment. We can always print copies from our flash drive. We do not want to risk losing any information by examining the tablet further. There may be more information in the disc that we are not able to recover from the scan."

Syed then switches to his lighter side. "Who knows how the disc is designed. It may self-destruct if we tamper with it; like in your *Mission Impossible* movies."

Syed's comment gets me started again. "Your mission, should you decide—"

"Do men ever grow up?" Linda asks, knowing the answer.

Syed and I both utter the same response simultaneously, "No."

I continue examining the documents, one by one, as Syed and Linda scoot closer to the computer screen. I immediately grasp the general thrust of some of the ideas presented while skipping over others where my powers of comprehension fail me. And then I see it—*my* work, *my* theory, full-blown and unvarnished, clearly not a work in progress.

"Where did this come from?" I ask, surprised, angry, and jealous all at once. "This is *my* work."

Linda peers in closely. "What is it? What do you mean *yours*?"

"Well, I've never seen this concept fleshed out, complete, but it follows from my work. Yes, and others; but I have had the lead on this for some time now."

Linda looks again at the document and then at me. "What is it? What are you working on?"

"In a nutshell, it has to do with the Big Bang and the origin of the universe. The essential question I have been attempting to answer is, 'What ignited or initiated the Big Bang?' In attempting an answer, I have followed on the work of colleagues who had proposed a hypothesis that the universe materialized from a black hole; itself residing in a higher-dimensional universe. The Big Bang, in this view, is nothing more than an illusion occasioned by a collapsing star in another universe vastly dissimilar to our own. As I said, I am not the only one working on this idea but how the theory is formalized in this document reminds me of my work." On a humble note, I add, "I could be mistaken, but I know my *signature*, and this smacks of it *right between the eyes*."

"Maybe it *is* yours," says Linda, in a lightning-quick response.

"I like what I'm hearing," I say. "Go on."

"If these documents reflect what will happen in the future, why couldn't you be the one to have eventually completed the work?"

We all have a good laugh, the mind-boggling absurdity hitting us like a cold shower.

"So, what's the point of my even finishing the work if I'm already looking at it?" I say, adding another layer.

Syed lets out a genuine belly laugh as Linda and I join in.

Contributing to our ribaldry, Syed carries the day: "Perhaps you should retire now; your work is apparently done for you!"

It takes us quite a while to settle down after that one. Our laughter releases the built-up tension we have accumulated, providing a welcome break from the intense concentration as we pour over the details of our find.

Another document outlines the black hole information paradox, resulting from quantum mechanics and general relativity—it proposes that information could disappear in a black hole, permanently, resulting in numerous physical states resolving into the same state. The proof debunks the notion out of hand.

A document detailing a "stable warp bubble" and "faster-than-light travel" goes far beyond previously published reports from NASA—electromagnetic propulsion drive becoming a reality.

And so it goes, many of the entries too opaque to analyze quickly.

The next one that stands out deals with information technology and surveillance. It reminds me of the book I read on the flight over but taking matters to a whole new level.

"Look at this," I say, drawing the attention of Syed and Linda back into the analysis. "It seems to be an algorithm, the granddaddy of all algorithms. You've heard of Big Brother? Meet his dad."

"I don't think I want to hear this," says Linda. "What are they up to now?"

"The algorithm appears to be directing surveillance technology. Nothing new here, but this diagram shows a 'net' blanketing all information and communication devices, collectively, in real-time. And here's the scary part—there is a built-in control process that can *hijack* every accessible communication device anywhere simultaneously and reconfigure the information after or even *during* input. The diagram isn't written in code; it is a clear and unambiguous schematic outlining, step-by-step, a procedure for controlling information technology on a universal scale."

Linda immediately grasps the far-reaching implications. "You mean the process could rewrite a text or email after or even *while* I'm typing?"

"You remember those times when you were having trouble with your computer and you called tech support?" I say, by way of analogy. "They sometimes ask for permission to direct your computer remotely to fix the problem. This is like that with three exceptions: they don't ask you for permission, they don't tell you what they intend to do, and there is no *they*—it's all automated by software, no human intervention is required. The program follows a universal template and there is no indication that it is limited to any particular nation or language. I can only speculate but I believe it is a program specifically designed to impose a conformity of ideas, an all-encompassing, ideological straightjacket."

Syed has been listening carefully. "How much capacity would you need for such a monster?"

"Yottabytes," I imagine.

"From the Jedi Master," says Linda.

"No, I don't think so; it's just a very, very . . . very large number."

"So much for Moore's Law," says Linda, and then, with a tinge of despair, she adds: "I don't think I want to know anymore."

"We're just looking at information," I say, putting things into perspective. "We don't know if any of this will be implemented."

"And that makes me feel better, how?" says Linda.

Syed moves the discussion forward and in another direction. "How are we to use this information? At this time, we are the only ones privileged with this material. That would give us some advantage, would it not?"

We look at each other to see who will blink first.

Linda breaks the silence. "But are we really in control of this information? I mean, what say do we have in what happens in the future? Can we do anything to speed up or impede how it unfolds?"

Syed interjects one of his cinematic metaphors. "We *have* the script, the documents. We can change the plot . . . the denouement. It is surely up to us to orchestrate the events. Why else would we have this information? Is it not meant for us?"

"You bring up an important metaphysical problem," I say. "Are we the recipients of a gift? Is this information meant for us? I don't know."

Linda can see that this line of inquiry isn't getting us anywhere. "I can't wait to see what the other tablets contain. I wonder what discoveries we'll find."

I know Linda is anxious to find out what the tablet containing the DNA marker might reveal. So am I. "What do you hope to find?" I ask.

"I don't know," says Linda. "We haven't even scratched the surface yet. What we've learned so far is surprising; it's also a little troubling, especially the document about surveillance. I can't imagine what the other tablets will reveal."

"It will take some time just to complete our examination of what we have," I say. "Some of these documents contain more information than I expected; it's all scrunched up. It slows the examination process down considerably having to stretch and pull the information on the documents to get the content." Feeling the strain getting to me I suggest we take a break.

"Yes," says Syed, "let us rest for a moment. You have more time to finish the analysis. We can do so later. Come, let us return to the living room. I will make some of your American coffee for you."

Linda and I sit on the sofa next to the large window overlooking the courtyard and wait for Syed to join us.

"Now that we know we can retrieve the information," says Linda, "we need to get the other tablets to the lab. Do you think Syed can set up another session soon?"

"He can get anything arranged if he wants to. He's holding the strings, remember?"

Linda smiles, looking more relaxed. It feels good to take a break. The mental effort has been draining.

"Thanks for your insights," says Linda. "It must have been taxing to sift through all of those documents."

"They say chess masters burn quite a few calories during a tournament. If that's true I must have burned my share mentally deconstructing all of that data."

Syed joins us.

Linda and I are delighted to have our American coffee—one of life's simple pleasures. Syed occasionally has coffee in the morning but prefers tea, with milk.

Linda takes a sip of coffee and asks Syed, casually, "When do you think we will be able to get the rest of the information in the other two tablets?"

"I will call and make the arrangements."

That's all Linda needs to hear. She knows she is on deck to interpret the information from the DNA tablet and is eager to participate.

"Your turn to burn calories," I say, kidding with Linda.

Syed looks puzzled. "What do you mean?"

"I was telling Linda that mental work burns calories like a chess master playing in a tournament."

The word chess strikes a chord. "Oh, I see," says Syed. "Do you play the game, Jake?"

"I have dabbled on occasion," I say, downplaying my interest and skill. Something tells me I will need every advantage should Dr. Azad offer a challenge.

"We should have a game sometime."

Syed's offhand delivery doesn't fool me. I suspect he would make a formidable opponent should I accept. "I'd like that," I reply, hoping he will let the matter lie.

I imagine how the game will play out—a calamitous defeat, my ego plummeting as Syed demolishes my minor and major pieces, one by one, followed by a "brilliant" but wasted sacrificial move on my part; and finally, banking on my pawns to carry the day, I watch helplessly as Syed slaughters them like so many weeds under the whirling stroke of a powerful trimmer. I then imagine Linda's chagrin as she witnesses the aftermath, her "hero" *dying* on the battlefield. Okay Jake, get a grip. I've already lost the chess match before the game even begins. It's the *inner game* that counts—the mental toughness you bring to the challenge. I can do anything if I set my mind to it! There . . . that's better.

"After completing a preliminary analysis of the remaining tablets," says Syed, "perhaps we should celebrate. I have been meaning to show you the view of the city from the Cairo Tower. It is very impressive."

I welcome Syed's invite; it should put a nice cap on what has proved to be a day of intense exploration.

By the time Syed retrieves the data from the remaining tablets, it is late afternoon.

We blithely accept a filtered view of reality, believing reports of news events with little or no scrutiny. We have gone from a world where the "news" was delivered in a mostly unbiased manner to one in which it is selected and pre-screened to determine whether or not it conforms to the reigning orthodoxy. Media outlets routinely promote, select, and edit coverage before public consumption. Misinformation, mendacity, and manipulation in the media have bled into other areas once considered sacrosanct and free of bias. The scientific community has not been immune. Issuance of grant money is often contingent, based more on adherence to political ideology than an unprejudiced search for the truth. Plagiarism and falsification of scientific data are not uncommon. How reliable is the information we currently possess? We have no idea where it came from.

Before we reconvene in the studio to begin our review of the DNA documents, I bring up the key question that has been occupying my mind: "Can we trust this information?"

"I wish you hadn't brought that up," says Linda. "I don't know how to answer it. You looked at the information. Didn't it make sense? I mean, I hope you're not suggesting it has somehow been fabricated. Are you?"

Linda seems frustrated but the question had to be asked. "No," I say, "just questioning everything; trying to be sure we've covered the bases."

Syed intercedes to balance out my argument. "Perhaps we can take a look, a quick review, of the data on the DNA tablet. That might give us a better perspective to judge the evidence; see if it provides further support for one opinion or the other."

"You could have had a great career in the state department," I say. I meant it as a compliment and hope Syed takes it as such.

Pulling out another of his stock phrases, Syed says: "That would be 'a fate worse than death.' "

We all take it in good humor. It is clear to us we should "leave no stone unturned" in our search.

Looking over at Linda, I say, "You're up."

Linda, ever the quick-study, tags the documents and sets everything in order without Syed's assistance.

The first few documents reveal nothing new, as we suspected. The first illustration depicted one of the works of Galen, dating from the second century A.D., showing a rudimentary diagram of the circulatory system. Why this particular representation and not another? Why Galen and not another scientist? It is too early in our investigation to say for sure. We might never know.

Making her way through several documents without comment, she stops at a record displaying a work she recognizes as being from Antonie van Leeuwenhoek, the "Father of Microbiology." It presents a diagram, a crude model of a microscope attributed to him.

Linda looks puzzled.

"What's so interesting about this one?" I ask.

"Leeuwenhoek didn't invent the microscope. Why would there be an exhibit of his 'improved' version? It doesn't seem right. It's not consistent."

I know better than to dismiss Linda's comments. She usually has a reason to home in on any detail that might bear further scrutiny. "Who invented the microscope?" I ask, wondering where Linda's suspicions might take us.

"Certainly not Leeuwenhoek," says Linda. "There were one or two others that preceded him. It's not important who they were. I don't even remember. I just remember that it wasn't Leeuwenhoek. I remember because, when I first learned of his work, I thought he *was* the inventor but my professor set me straight on that and I have remembered it ever since."

I can see Linda is stuck. She sits there staring at the screen, forehead wrinkled, concentrating.

"Maybe it doesn't mean anything," I say. "The problem is we don't have anyone to ask. This whole exercise is one-sided. We're pretty much on our own." Oh boy, that was helpful. I can do better. "Maybe his contribution was more important. Maybe it was his work that brought the instrument's use to better benefit for others." That sounds better.

"I guess that makes sense," says Linda.

Whether it did or not doesn't seem to matter. Linda continues her search. After briefly reviewing numerous additional documents—working through everything from the carbon cycle, epigenetics, the smallpox vaccination, meiosis, blood types, and so on—we come to item fifty-one, the double-stranded polymer molecule—DNA.

Syed has so far remained silent while Linda checks each item. He then jumps in with an interesting question: "The discovery dates of these 'marker' plates, do they match? Might they be coordinated in some way? I do not know if this would tell us anything . . ."

I am curious about what Syed has in mind with this line of reasoning. He doesn't seem so sure himself. "The Schrödinger equation was published in the nineteen-twenties," I say. "Watson and Crick did their work in the early fifties so—"

"That's true," says Linda. "But DNA itself was known back in the nineteenth century. They didn't understand its genetic properties back then but, either way, we look at it, I don't see any match here."

Syed once again puts things in perspective. "You are right. I do not think we can make so much out of the markers. Perhaps they are serving to identify a class of data. Each represents some part of a whole. It is not clear."

"There may be another explanation," I offer. "The Schrödinger equation plays stand-in for Newtonian laws and energy conservation in classical mechanics. The wave equation allows us to predict the probability of events and the future behavior of dynamic systems. It is integral for analyzing and calculating quantum mechanical behavior. It could be considered the linchpin of modern science. There are other methods for making predictions in quantum mechanics; the path integral formulation or matrix mechanics, for example. But I believe the Schrödinger equation has even greater applications than currently in use and may even lead to a greater understanding of phenomena in other fields outside of physics as well. There is already evidence supporting this view. Many of the newer discoveries documented in the plate make extensive use of the equation. Maybe the same thing can be said of the DNA tablet as well. What do you think Linda?"

"From what I have seen so far DNA seems to be playing a similar role. And we haven't completed our review of the first tablet or even begun analysis on the last one yet. Once we complete our review of all the tablets, we should be able to confirm how important the 'markers' are in our findings."

I appreciate Linda's reminder, in so many words, that we are scientists. As scientists, it is our responsibility to examine *all* the evidence before drawing any substantive conclusions.

"Let's keep reviewing a while longer," says Linda. "I want to see where the information will take us."

With that, Linda continues the review and analysis, skipping ahead to—as she puts it—"Get to the good stuff."

"I've heard about this," Linda says. "I wish I had gotten into this area. Oh well, it's too late now."

"What are you looking at?" I say, doing my best to follow along.

"It has to do with the connection between genes and the brain. The schematic traces the connection between a specific gene variant and the dorsolateral prefrontal cortex. The connection had already been established; that's the 'work in progress' part. But now it looks like we've been able to either introduce the variant in people without a copy of the genetic variant or 'knock out' one of the genes in those who have two copies; since it is the single-copy variant that promotes—and get this—both longevity and brainpower. We are now able to slow down the natural aging of the brain. So, we've taken an idea from its infancy and brought it to market, so to speak."

Syed hasn't said very much during our review but when he does offer his opinion, it is spot-on: "That would have even broader implications as I see it," he says. " If we have learned how to do the procedure in this case, would it not apply to others as well?"

"That's right," says Linda. "That would mean we have figured out how to manipulate our genome in a very significant way. Can you imagine the ethical problems, not to mention the regulatory nightmare in getting such a procedure approved? Something must have happened to open this door, some kind of emergency. I wonder if . . . I don't know; it's too much to think about right now. And after your 'Big Brother'—make that 'dad' of Big brother—bombshell, I'm afraid to look at the next document."

"Nothing to worry about," I say. "The next trick up their sleeve is probably just a rudimentary stab at mind control. They'll just invade our neurons and plant whatever they want into our brains and control us remotely from the president's hideaway in *Raven Rock Mountain*."

"Now I'm even more afraid to look. Thanks, Jake. Thanks a lot."

Linda pulls up the next document. "So, they've finally done something useful with nanobots. Cool."

"That's not fair," I say. "They've made some progress—"

"Okay," says Linda. "I was just kidding. But this is really something. Remote-controlled medicine is now fully online. Wow! Biocompatible fuels; the whole nine yards. I've read about this and now, here it is—a done deal. Incredible."

Continuing, Linda makes passing comments as she wades through the documents: "Bioelectronic medicine, check. 'Junk' DNA . . . huh, I thought so. Cockroaches? Leave it to them to bud in—you can't get rid of them. Well, at least they were useful . . ."

Linda pauses for a moment and then brings up a strange idea. "It's like I'm in the future looking back in time when all of these ideas were incubating. We *are* here now, right?"

I hadn't followed half of what Linda was muttering, mostly to herself. And I doubt that Syed had either. As far as the "cockroaches" comment, there is no way I'm touching that one. But now Linda is raising the possibility that we are actually *in* the future. As General McAuliffe said when asked to surrender during the Battle of the Bulge in World War Two, "Aw, Nuts!" Is Linda losing it?

"Okay, now you're scaring me," I say. "What are you *talking* about?" I'm not sure if Linda is serious or just exploring possibilities. Either way, it doesn't sit too well with my paranoid, schizophrenic, and conspiratorial mindset.

"Well, you said yourself that one of the concepts in the tablet you were reviewing had your 'signature' on it, to use your word. Maybe all of this information, all of these 'discoveries' we're looking at, is history. We haven't checked. Maybe we had a mental blackout and—"

Syed jumps in: "We should take a break," he says, trying to settle things down. He looks worried, concerned. "It is too much. Why don't we put everything on hold for now? We can finish later. I promised you a visit to the Cairo Tower. It will be dark soon and we can enjoy the view of the city. Come, I will take us. There is no need to call Emile. The traffic should not be so bad. It is a short drive."

"You're right, Syed," says Linda. "I am jumping to conclusions—"

"You've been hanging around me too long," I say, in an attempt to ease the tension and help Linda save face after her bizarre comment.

"The 'Jake virus' " she says. "It's highly contagious!"

I am relieved Linda had only taken herself half-seriously. We could use another respite anyway. Too much mental drudge work can tax the brain to where you're no longer thinking straight. If it can happen to Linda, it can happen to anyone, especially me.

The Cairo Tower is the perfect spot to view the city. The lotus plant-like structure covering the tower, as Syed explains, is symbolic—representing the plant used to make papyrus. We go up to the VIP restaurant overlooking the city, Syed having customarily arranged an inside track circumventing the lines. It is approaching twilight and the scene is even more spectacular than we expected. Our trip over to the Cairo Tower took less than ten minutes, of course. I am by now convinced that Syed's residence, besides being stylish and meeting his needs, is strategically located for his benefit.

Syed recommends we try the calamari, fried squid. "It is very good."

"I'm afraid not," I say. "I tried it once, and only once, at a restaurant in San Francisco. I took one bite and asked the maître d' if I could exchange it for something else. He was gracious as I remember. It's just not for me. But don't let me stop you. Go ahead and enjoy it. I think I'll try the pasta; it's safer."

Syed orders calamari and Linda fish and prawns in white sauce.

Linda and I play it safe and ask for tea while Syed defers. "I like to have a little cup of coffee, or what we also call ahwa, after the meal," he says.

I hadn't planned on asking but I can't help it; I have to know. "Linda, I know this may not be the best time to bring it up but what was that comment you made about 'cockroaches'?"

"Oh that," says Linda, like it's no big deal. "Biophysicists have harnessed a roach's ability to assemble light signals from their photoreceptors over time and use the accumulated signals to see in the dark. Their research has resulted in a substantial improvement in night-vision technology. That's what I got out of it. I may have missed something."

"Sounds like that little slice of information should have been in my—I mean our—plate dealing more with physics than genetics or biology. I guess we could split the difference."

Linda takes the question seriously. "The diagram shows a connection between a roach's nervous system and neural signals, so that might explain it."

"Roaches are better for snacking," says Syed. "China considers them a delicacy but I have never tried them personally."

Now, this is "guy talk" I can relate to. "Perhaps with a little marinade sauce just to—"

"Stop it!" interrupts Linda. "Enough of that."

"Yes, ma'am," I say, acting the part of a scolded child.

Now is a good time to change the subject and Syed loses no time in doing so. "How are you enjoying the view?" he asks. "Isn't it . . . what is the right word . . . panoramic?"

"Yes, it's breathtaking," Linda says, taking in the view. "How high up are we?"

"The tower is about a hundred and eighty meters or so. Oh wait, that would be in feet. How much is that? Maybe close to six hundred. It is the tallest structure in Cairo. You can see the Grand Hyatt just over there," he says, directing our attention.

We finish dinner and return to Syed's home. Linda excuses herself and retires for the evening. "I think I'll go to my room and read for a bit," she says.

"You're not going to leave me along with Syed, are you?" I say. "He'll probably challenge me to another game of billiards so he can humiliate me again."

"Goodnight gentlemen," says Linda, intentionally ignoring my less than witty riposte.

"This would be the perfect time for a game. What do you say, Jake?"

"Not billiards," I say. "I learned my lesson."

"No, no I mean chess. Are you prepared for a game?"

"Sure, I'd like that." Too late now; my brain spoke before asking me.

"I will set everything up in the game room. Come and join me. It will only take a moment."

And so it did. Syed takes the board out of the cabinet where he keeps the game of Senet. All of the pieces are already in place and ready for action. I get a sinking feeling.

Syed sets the board down on the table with the white pieces in front of me.

"Aren't we going to choose to see who gets white?" I ask. It is the gentlemanly thing to do.

"No, you are my guest. It is of no consequence, please."

I began with my usual opening, *Ruy Lopez*—the *Spanish Game*. Syed counters with knight to f-6, subtle. Now I must either defend my pawn . . .

My mind immediately goes back to the basics. Playing against Syed is intimidating and my only mental defense is to force myself—not always a good strategy; let the mind run on automatic they say—to remember: develop quickly, castle early, knights before bishops, control the center, don't do anything dumb (there are no take-backs); and, most importantly, don't get "forked"!

Development goes evenly, neither Syed nor I gaining an advantage, and, by the middlegame, it remains anyone's game. By the time we reach the endgame neither of us has gained an edge. Finally—Syed with king and bishop (on the white square); me with king and rook—it is obvious we have reached an impasse. We agree to a draw. I feel the same way as I did back in grade school when the intercom announced, loud and clear, the next victim to be called to the principal's office . . . and it wasn't me!

"Good game," says Syed crisply, adding: "As the British say, 'jolly good show, old chap.' "

"Here, here" I counter in my best British accent.

If Syed knew I had just played the best chess game of my life, he may not have been so generous. Note to self: Avoid a rematch!

"Well," I say, "shall we call it a night?"

"Sleep well, my friend."

Chapter Fourteen: Keeping Track

Syed is in the studio when I come downstairs. I join him and am surprised to see that he is already screening the last tablet.

"Good morning," I say. "I see you have a head start. What are you looking at?"

"Hello, good morning. I decided to see what information the final plate might give us. I hope I am not getting too much ahead. I thought—"

"No, I'm glad you're taking the initiative. We can easily catch up. You can just give us a briefing. Linda should be with us shortly."

"She is quite intelligent, you know."

I am surprised to hear Syed speak for the first time about Linda outside of a general and professional context. His comment demonstrates his appreciation for her contribution as well as her intellect.

"She continues to surprise me," I say. "I knew she was smart the moment I met her but I have since discovered that she is much quicker than I will ever be at grasping the kernel of an argument or problem and just as quick to come up with a possible solution."

Syed is just about to comment further when Linda comes in and joins us.

"I see you've started without me," she says, affably.

"No, not at all," says Syed. "We were just bringing up the data to prepare everything in advance."

Not perfectly accurate but it hit the right diplomatic cord.

Being intelligent has its pros and cons like anything else. I remember back to when I was fifteen-years-old and participating, along with many others my age, in a three-day process designed to measure intelligence. I was not privy to the results (they were considered "proprietary") and, when I asked about the findings of my assessment, I was told that it was none of my business, in so many words. Being the clever young boy that I was I managed to steal a peek at the results but, just as I was getting to the good part, the file was yanked from my hands and I received a stern warning from the assessment administrator. I did however read far enough into the document to learn that my I.Q. was "not tractable." I didn't get to the part explaining the reason for my *intractability* but I knew what it meant.

After that little experiment—for the next several days—administrators, counselors, and academics bothered me endlessly until, finally, they left me alone.

So, I'm smart, so what. Lots of people are smart. So was the Unabomber; so were a lot of people that either misused their gift or were never heard from. The character Forest Gump, in the movie bearing his name, got it right: "Stupid is as stupid does."

The same does not necessarily apply the other way around. Intelligent people can do the stupidest things sometimes. I know; I'm one of them. We all accepted the preposterous notion that information can be stored in nothing more than a "rock." How dumb is that? Why were three purportedly intelligent people so easily fooled? We believed because we wanted to believe. I can't speak for Syed and Linda but I am determined to raise my game.

"Go ahead, Syed. You're in the driver's seat."

Syed catches my meaning. "I will do my best to steer us to our destination."

"Would you mind if we did something a bit different this time?" says Linda.

"What do you have in mind?" I ask.

"Well, I'd like to take a look at the last plate first. It might provide a clue to the timeline; how far out these plates document the future. Just a hunch."

Syed pulls up the last document in the third tablet. "It looks like a star chart," he says.

"Incoming!" I say, instantly regretting my flippant comment.

Linda tactfully ignores my remark. "Why would we want to see a star chart?" Looking closer, "It looks like a shift. Is that what you're seeing?"

"Without a doubt," I say. "But what is it exactly? It looks like an object headed toward us but shifted from one trajectory to another."

"And another," says Linda. "There are a series of shifts if I'm reading the chart correctly. We may need some outside help with this one. I don't know how to figure out what it might mean exactly."

Syed agrees with Linda that it is difficult to say exactly what it might mean. "We would not have this chart if it were not important. Looking at the chart without any reference is not telling us everything we need to know. There is no explanation, just the chart. I agree that we may indeed need some help."

"Is there anyone you can think of at the university that might help us?" I ask.

"Yes, but we would have to be discreet. I am sure Professor Moustafa would be of assistance to us."

"Well we're not experts in star charts," Linda offers, "but I think we can figure it out. There's got to be enough information on the net to give us some clues . . . Don't you think?"

"Yes, we could do that," Syed replies. "However, we can never be sure we have done so correctly. We will need confirmation to be certain."

"Well," I say, playing the role of mediator, "why don't we see how far we can get, see if it makes any sense, and—if push comes to shove—take it to the University for verification."

"Now Jake, there will be no pushing and shoving."

I'm not sure if Syed had taken me seriously or was just joking. It doesn't matter. I want to get to the bottom of this as soon as possible.

"We have to first figure out what this object is," I say. "This chart shows the ecliptic plane and an object approaching the inner solar system undergoing three separate shifts in trajectory, which makes no sense. It looks like the diagram is depicting the original trajectory and two adjustments. Is that what it looks like to you?"

"That seems right," says Linda, Syed nodding in agreement.

"The problem," I say, "is how to identify what object we are looking at. I assume it must be a comet or asteroid. What else could it be? The last shift—the one on the inside that cuts into the Earth—that can't be the original trajectory, that wouldn't make any sense. We would know about it already and be taking countermeasures to try and prevent or lessen its impact. Besides, this diagram wouldn't even be here if it was old news. It's the *last* entry in the log . . . in the disc."

"But as you said, we need to narrow it down," says Linda, getting drawn into the pursuit. "We need to know *which* comet or asteroid . . . or whatever it is we're examining. We need to track the trajectory. Wait a minute. Let me check a tracking site. I know they have one."

Linda jumps on the net and goes to the NASA website and pulls up the "orbit diagrams" section. "Oh, this is great," she says, frustrated. "Now all we need is the 'Object Number, Designation, or Name.' Perfect! If we *already know* what we're looking for, it's easy to find. Oh, and look at this," she continues, growing even more frustrated, "they've conveniently provided a laundry list of almost sixteen hundred 'Potentially Hazardous Asteroids' for us to skim through. That should narrow it down. It's hopeless."

This isn't like Linda. She's usually tough as nails, always up for a challenge. All of this must be getting to her.

"Well, what are our options?" I ask. "There has to be a way to narrow our search. Although I admit I'm not sure what it is right now." I am trying, unsuccessfully, to contribute to a solution but missing the mark. "Isn't there some way to overlay this diagram with other known plots to get a match? You know, the way they match fingerprints. There must be some program—"

Syed senses our growing frustration. "The Astronomy Department should have access to such a program but there is the problem of scale. If the diagram we have is not to scale, there is little possibility of matching it perfectly."

Linda looks hopeful. "That's a good point, Syed. I would think the chart would be to scale. So far, we've been handed everything on a silver platter. I don't see why this chart would be an exception. It's worth a try."

With that, we decide to pull out our ace in the hole and take a copy of the diagram to the university.

"That would be Professor Moustafa?" I confirm.

"Yes," Syed replies. "He should be able to assist us."

"So, what's our story?" asks Linda.

We know what Linda means. If we are going to get help, we have to have a cover, some innocuous tale to explain why we need the document analyzed.

"I got an idea," I say. "I could just say I was visiting Dr. Azad and had some papers in my briefcase. You know, some student assignment I handed out . . . No, that won't fly. That's too dumb. I don't know. You're clever Linda. You think of something."

"You could just tell him the truth . . . Well, sort of."

"This I gotta hear," I say.

"Sure; just say you came across the document in confidence. You don't know exactly what it might mean but were told it was important and the person who gave it to you couldn't tell you anything else. I know it sounds cryptic but that's the hook. Professor Mustafa will be glad to solve a riddle. You both have excellent reputations. Even if it sounds a bit out there, he'll cut you some slack. I don't have any other ideas. It's worth a shot. What's the worst that can happen?"

"Well, for starters," I say, "he might think we're nut jobs and call the—"

Linda and Syed laugh.

Syed seems hesitant to commit to this harebrained idea. "I have another idea," he says. "Perhaps it is not so difficult. I could call Professor Moustafa and ask him if he knows of such a program. I am sure he wouldn't mind loaning it to us or allowing us to make use of it in his lab. He may be a bit—what is it you say?—nosey, but we can work our way around that problem easily enough."

I have to admit it sounds better than either Linda's or my idea. But I'm not saying anything. Linda and I have been getting along splendidly and I don't dare say anything that she might take as an insult.

"That sounds less complicated," says Linda, graciously.

Syed calls Cairo University and asks for Professor Moustafa. His class is in session so Syed leaves him a message.

We decide to leave things where they are for now and wait to hear from Syed's colleague. Syed suggests we sit in the back courtyard and have our usual morning brew. I thought he would never ask.

After we get comfortable the talk returns to the document we have been reviewing.

"Why didn't they just tell us?" I ask.

"What do you mean?" says Linda.

"Well, here we are mentally toiling away looking at an image of an incoming object. We don't know what it is and we're doing our best to find out. Meanwhile, all we needed was a tag, something to identify the object, to tell us what we're looking at . . . and we get nothing; no marker of any kind. How could that happen? Up to this point, everything we've been reviewing is at least understandable. It might take a little work on some of the documents, but the information we need is there."

Syed has been following everything I have been saying and then, with a simple statement, solves the mystery: "Maybe they made a mistake."

Linda and I look at each other. There is a brief silence . . . and then we all burst into laughter.

Riding the good humor generated by Syed's penetrating analysis, I say: "By George, I think you've cracked the case!"

Syed reciprocates: "Elementary, my dear Banner."

Picking up on the Holmes reference, Linda adds: "When all else fails . . ."

"Unfortunately," I say, "we're still stuck with our problem. There are sophisticated programs that can decipher these types of diagrams. It depends on what Professor Moustafa has to work with."

"What if the professor isn't able to help us?" Linda enquires. "We need a back-up plan."

"I learned Fortran as an undergrad to avoid having to rely on others to write the code I need for research. It would take some time but I could write the code if I had to, as a last-ditch effort. But I'm sure there are already programs out there that would work."

"I thought using 'C' is more in vogue," says Linda.

"I just stick with what I learned. They both get the job done. I got comfortable with 'looping.' "

"That figures," says Linda. "You've always been loopy."

Syed isn't following our conversation so I let Linda have the last laugh.

Syed's phone rings. It is Professor Moustafa. Linda and I listen to Syed as he speaks in Egyptian Arabic, comprehending nothing. I should at least learn a little Arabic. I am in a Muslim country after all. Maybe Syed could . . . focus Jake . . . focus.

Syed finishes his call with Moustafa. "He says it would be better if I could send him the diagram. Transferring the program would be too much trouble. Using the program correctly would also pose a problem since we are not trained. He was also concerned about getting into trouble with the administration if he should do so without first getting permission. I told him I would call him back shortly. He has some time in his schedule before he must get back to his lectures."

"Well we can't doctor the chart," says Linda. "That would look messy and interfere with his efforts. What would happen if we sent it to him the way it is? Do we have any choice?"

Syed understands the situation completely. "He will of course realize what he is seeing. It is no secret. The diagram shows the original path and it is clear from the chart that the path, the direction, has changed significantly. If he can identify the object, he will be wondering what is going on. And he will certainly have reason to be concerned. Of this I am certain. He would not expect me to be sending him something that is of no consequence, for no serious reason."

"I just don't think there is enough to go on," I say. "But maybe we can figure this out without the professor's help. I'm sure I could use the computer system back at Stanford. No one would have to know. I have clearance to use almost any system I need. It's not as if the information is classified; it's just a matter of finding the right program. Once I'm able to do that I can scan the chart and upload it . . . No, that won't work either; too much trouble."

Then another idea hits me. I don't know why it hadn't occurred to me earlier. Too much information in my head competing for attention, I guess. "Syed, we could use your computer. It would take quite a few technical steps but it could be done. We could transfer the program we need . . . No, never mind. Besides, we have the tools we need right here. We just have to figure out a way to get this done without giving anything away. How do we do that? Back to square one. Sorry, just thinking out loud. I'm sure we're missing something. There's probably an easier way to—"

"Is there any possibility we can use the computer programs available at Cairo University?" Linda asks. "If we go to your office . . . No, wait . . . Syed, you can access the files you use at the university on your home computer, right?"

"Yes."

"Do you have system-wide access or is it limited to your work?"

"Well, I don't know," says Syed. "I can look at everything I need. I have the internet as well, but we have this at home. That is a good question . . . let me think. I can log in to my work computer and see if we can search for other programs. It may possible. Let us see what we can find."

We go back to Syed's computer and he logs in under his administrative password.

"Now where do we go?" asks Syed, speaking his thoughts out loud. "Let me see . . . here is a list of some of the programs. This one looks promising."

Syed attempts to open up a program under the heading: Astronomy Department. There is a prompt for a password.

"What is this doing here?" Syed mutters. "We are not searching for military secrets."

Syed enters his password and it is accepted. "I see I have more authority than I have been using. I wonder what else I can . . . No, that is not important."

Syed manages to access a program that he thinks might help us.

"Look," says Linda. "There's a process right here. Click on that."

Linda has zeroed in on a procedure that allows us to look at actual diagrams without having to first identify their designation criteria. After we figure out how the information is organized and look at a few of the formats, ignoring what we agree is irrelevant, we land on a process that allows us to overlay schematics for comparison. The charts we pull up all have data points marked on each diagram. The program is quite clever. It has a universal application that accepts schematics from multiple sources and reconfigures outliers to conform to the program's format. Our diagram doesn't have any data points but, as far as we can determine, all we need to do is upload the chart and the program will do the rest. We scan the chart and, after some difficulty, complete transferring the chart onto the computer interface, a holding platform. We still need to transfer the diagram to the actual template for analysis. To do this without data points we need to overlay our chart onto an existing chart. Any will do. We pull up an arbitrary chart showing an incoming asteroid that *does* contain data points. The program allows us to transfer the chart, superimposing it onto the existing "dummy" diagram and—bingo! The data points appear as if by magic!

"Whose program is this?" I ask. "This is fantastic."

Linda keeps the chart open and scrolls up to the top where the architect of the program is identified: Stanford University.

"Synchronicity!" I say. " 'I love it when a plan comes together.' "

"I didn't think we were taking the lead in astronomy," says Linda.

"Could be some postdoc from who knows where," I say. "What matters is that it's just what we need right now."

"Why wouldn't Moustafa know about this program?"

"He probably does know," says Syed.

Linda lets Syed's remark pass without responding. No sense our getting involved in internal politics.

Syed takes his turn to get us back on track. "What is the chart telling us? What is the object we are seeing?"

Linda scrolls back down to get a full view of the chart. We look at the path of the object on the diagram and it is—no surprise—identical to the original. We then look for the proper designation to confirm the identification. The orbital path is highlighted in blue and the designation: "(1997 XF11)."

"What are we supposed to make of this?" Linda asks. "This looks like . . . well, that's the original path. But wait a minute. We can't tell anything from looking at this chart. I mean, we know now what object we're dealing with but how can that help? Let me look somewhere else. This isn't giving us everything."

Linda pulls up the information on the asteroid. "It says here that in twenty twenty-eight the asteroid will come extremely close but will miss the Earth. It says if it hit the Earth it would be a mile-wide asteroid striking us at about thirty thousand miles per hour. It would have energy approximately equal to a one million megaton bomb. And get this: it says the asteroid would wipe out 'most of the life on the planet.' It would be an 'ELE' or 'Extinction Level Event.'"

Linda's analytical mind goes to work: "So what we are calling 'shifts' in the asteroid's incoming flight path may be due to updates in tracking or—"

"It may have been deflected," I say.

Syed's usually calm demeanor gives way as he joins in the discussion. "What might cause such a 'deflection'? Are you suggesting the asteroid changed its course due to some kind of collision?"

"Possibly," I say. "If it encountered an asteroid or underwent some unknown non-gravitational forces."

"How big is this asteroid?" asks Syed.

Linda scours the site for more information. "It looks like it's between one-point-three and two-point-eight kilometers in diameter." She then pulls up another site. "Hey, look at this. Wouldn't you know it? YouTube has a mock-up video of an asteroid's expected impact on the Earth. Do you want to see it? It's a different asteroid but it might give us some idea about what to expect. It's less than a third the size of what we're looking at but—"

"Why not," I say. "It's probably grossly exaggerated, but go ahead, put it on."

We watch the six-and-a-half-minute YouTube video showing cataclysmic devastation and worldwide destruction.

"We don't know what the consequences might be from an impact," says Linda. "Who knows what it's made of or how compact it is. It could break-up in the Earth's atmosphere or hit us with everything it's got." Then, having second thoughts, she adds, "Or it could just be information."

"How do you mean?" I say.

"We don't know if it will strike the Earth. By the time it comes our way, we could send up some type of spacecraft or something to intercept it; blow it up or at least break it into smaller pieces where the damage of impact would be lessened."

"Well now you've taken the fun out of everything," I say. My understanding of physics tells me breaking it up wouldn't diminish the risk by much, if any. Better to deflect it if possible. She does have a point however; it doesn't have to be Armageddon. "You're probably right . . . but still; why is it the *last* entry?"

Syed has been following our discussion and comes to a different conclusion. "There may be another possibility for the chart being the last entry. Perhaps we were unsuccessful in our attempt to intercept it before impact."

"You could be right Syed," says Linda. "I didn't think of that. Just because we *can* intercept the asteroid, doesn't mean we did. Something could have gone wrong. *Failure* is an option."

"Let's take a closer look," I say. "There must be something we're missing."

"This might be something," says Linda. "The spacing that shows the shifting trajectories narrows as it comes closer to Earth. That could mean something."

"If they are time intervals," I conjecture, "that might mean the asteroid is accelerating, gaining speed. What could cause that?"

"I don't know," says Linda, "but that would definitely increase the impact."

"F equals ma," I say. "Force equals mass times acceleration—"

Syed is quick to draw the appropriate conclusion: "The greater the acceleration, the greater the force; the greater the force, the greater the impact."

"Exactly," says Linda. "It could be quite a lollapalooza."

"How do we even talk about these events?" I ask, more perplexed and confused than ever. "They've already happened? They are *yet* to happen?" Looking at Linda, I point out how she spoke of the possibility of *preventing* the asteroid's impact as if we were ahead of the event. It seems so; and yet, we have information in front of us suggesting it has already happened or is even imminent.

"Look," I say, "these documents reveal information and knowledge of future events. That seems clear to me. If that is true—and the evidence is overwhelming—do we have any say in what happens next? Is it even possible for us to act on this information and change the future? I know that sounds strange: We are here in the present, being presented with information from the future, and asking ourselves whether or not we can change the outcome. Ordinarily, the answer would be simple—of course, we can affect the future; it hasn't happened yet! In a nutshell, we've been presented with a fait accompli and we are asking ourselves if it is possible to reverse what has *already* occurred." Realizing the absurdity of what I have just said, I ask: "Did I state that correctly, or am I missing something?"

Leave it to Linda to set matters straight: "We're scientists. We are forgetting something—we haven't examined all of the evidence. We are attempting to draw conclusions and determine a course of action without taking into account that the answers we are looking for may be right in front of us. We need to complete the work we started."

"I agree," says Syed. "We may find what we need if we keep looking."

There is only one thing I can think of to say: "Let's do it!"

Chapter Fifteen: The Ant Swallows the Elephant

We have examined the first two tablets, though not in detail, and made a preliminary review of the third. The final tablet containing the cosmological marker needs further analysis before we can consider our initial run-through complete. Our goal: To discover crucial information explaining or tying together the three tablets, providing an answer as to why these tablets were preserved for subsequent discovery.

We are not interested in conducting an in-depth analysis of the documents; we are looking for a clue, a link—anything—that can tell us why these tablets contain the information they do and how they are connected. There must be an underlying theory or principle that holds everything together.

There are documents providing the final decimal points for the cosmological and galaxy power spectrums; an interesting (but unessential for our purposes) resolution to the black hole firewall controversy advanced in hopes of providing a possible solution to an apparent inconsistency in black hole complementarity; and many others that do not warrant further review. To save time I don't bother explaining the mechanics of these and other ideas, preferring instead to keep searching until I hit pay dirt. Neither Syed nor Linda object.

I keep searching until I come upon a document that seems to satisfy our search criteria. It outlines a theory that encompassed physics, genetics, and cosmology.

"It has been staring us in the face all along and we didn't see it," I say. "You clear away the 'quantum foam' and the cosmic haze and it stands out in stark relief." I can't help waxing poetic; it is marvelous in its simple complexity.

"Is this it Jake? Is this what we've been looking for?" Linda's dark blue eyes are riveted to the screen.

Syed is frozen in concentration, straining to see, to understand.

"This is it," I say. "This is what we have been looking for." And then, with absolute conviction, I say: "This is the reason we were born—to understand, to know . . . everything."

"Well what is it?" asks Linda, exasperated. "Don't keep us in suspense."

"Yes Jake," says Syed. "Tell us what you have discovered."

"They are related," I say, "in every way possible or imaginable—physics, genetics, and cosmology. Just as General Relativity subsumed Newtonian gravity, quantum mechanics has done the same to General Relativity. And there's more . . . much more. A deeper understanding of genetics is inextricably tied to quantum mechanical behavior and quantum mechanics, in turn, drives our understanding of cosmology, and finally; to top it all off, the Schrödinger equation underpins everything—and not in a general sense, in every sense. These mathematical structures, incorporating the Schrödinger equation, determine—not only the genetic coding of DNA—but the 'genetic space' in which it can operate. The equation also explains, in great detail, how we can 'scale up' from the quantum level to incorporate cosmological expansion via the cosmic background radiation. Not the standard 'Wilkinson' map delineating the isotropy of the cosmic microwave background or even a marriage of the power spectrums, but employing 'twenty-one-centimeter tomography,' a measuring technique giving us a 'new and improved' map based on quantum physics and making heavy use of, once again, the Schrödinger equation. The use of the term 'scale-up' is not strictly accurate, but words fail me in describing the nature and scope of these findings. In essence," I conclude, "we have two blueprints—the genetic and cosmological—both of which are derived from and understood by the use of the Schrödinger equation."

Linda and Syed are silent for a moment, absorbing the import of what I have been telling them. Then Syed asks a question that brings it all together: "Is this the end of knowledge?"

"No," I say. "It means that all future knowledge, everything that is knowable, everything that we can discover, is accessible to us. We no longer have to grope in the dark to discover the means to acquire knowledge. We now have the necessary *tools*. There is one constraint however, one limitation: These tools can only tell us what is knowable. What is beyond the reach of these tools will remain forever inaccessible."

Linda puts forward an appropriate analogy, summing up: "Maxwell's demon on steroids."

"This is a treasure trove," I say. "Previously unanswered questions are now accessible to us. Imagine the possibilities."

"And questions of logic and philosophy?" asks Syed.

"Those would remain open," I reply. "They are by definition beyond the reach of the physical sciences. Logic, axiomatic proofs, will continue to require mental effort and their resolution subject to review or amendment. As far as philosophy is concerned . . . well, philosophy is just that—philosophy. We will always argue the merits of ideas; what things mean, whether or not they have value or contribute to our well-being—what is the best way to live our lives."

"And we will always have our mathematical puzzles," says Linda. "The Riemann hypothesis and so on."

"Still," I say, "we now have the tools we need and that opens up endless possibilities. We are now able to resolve many outstanding issues that we have been wrestling with for some time; the nature and composition of the universe, for starters—"

"That's right," says Linda, grasping the full impact of what we have discovered. "We should be able to figure out the nature of 'dark matter' and 'dark energy.' Maybe there's already information in the documents about this."

"Before we continue," says Syed, "perhaps it would be best to consider what we should do about our discovery. How we proceed from here."

"That's right," agrees Linda. "We've been ignoring this question for too long. We need to figure out what to do next."

"Let's start with what we know," I say. "We have the information. Regardless of how we have come into its possession, it is now ours to use as we decide. We can retain possession and personally manage its application, essentially taking credit for what we have discovered. Or, we can make an announcement and share what we have with the world."

"Wouldn't that accelerate our knowledge 'ahead' of schedule?" says Linda.

"Yes and no," I say. "It depends, once again, on how we view the discovery. If we look at this information as existing in the present—which we have now confirmed—we have every right, even obligation, to make it available to benefit the world community. How could we not?"

"And if it is from the future," Syed reasons, "has it not been given to us to decide what we should do?"

"How long would it take to fully comprehend everything we have?" says Linda. "Even with help from others, it will take some time to fully understand the information and appreciate its full impact."

"Well, let's think about this for a minute," I say. "If we call a press conference and announce our discovery to the world, the question of ownership will come up. Whose documents are they? Who has the original claim to their contents? Us? And if so, what does that say about the original authors of these documents, the scientists who will make these discoveries in the first place."

"We don't know who they are," says Linda. "There's no indication of authorship on any of the documents."

"And there is still the matter of the last document in the last tablet," says Syed. "If we provide this information, it would cause a panic, would it not?"

These concerns assault my brain, demanding answers. Why is it up to us to decide? "If we release the information, as Linda says, we will be speeding up the future; if that makes any sense. It doesn't to me."

"I think we are obligated to make the information available," says Linda.

Syed concurs.

I also agree. "I think Mark Twain said it best: 'When in doubt tell the truth . . .' "

"I don't know about our enemies," says Linda, "but it will certainly 'astound' our friends."

Syed seems pleased with our decision: "It is settled then. We must make plans."

"And the document charting the asteroid?" Linda reminds us.

"It's all or nothing," I say.

No response. I take silence as confirmation. "Let's decide when and how we're going to do this."

We spend the rest of the day working out a timeline and other details for releasing the information. We still need to decide which documents to unveil during the presentation. We don't think it would be a good idea to inundate our audience with too many details. Everyone will get to see everything soon enough. The chart showing the incoming asteroid would best be kept under wraps for now. This isn't the forum to create panic. We also agree to avoid any arguments over proprietary claims and to publish the information on the World Wide Web after our news conference.

Ordinarily, I have little trouble sleeping. Tonight is an exception. I'm restless. I manage to fall asleep but wake up in the middle of the night. I get out of bed and walk out onto the balcony. I look up at the stars and wonder: The light from the stars in the heavens now reaching me contains information of events long ago; some reaching back to the time of King Tut. If King Tut, at the tender young age of eighteen, had peered into the night sky at this same spot over thirty-three centuries ago, the light reaching him would have begun its journey before recorded human history.

Light contains information, capturing events and carrying them across space and time. How is this possible? Where, specifically, does this information reside within the photon? When this distant light reaches me and impinges my retina, traveling along the optic nerve reaching my visual processing center in the *back* of my brain, I *see* history—prerecorded history. Conversely, I wonder, will an advanced civilization far out in space—in proximity to the stars I now see—eventually witness me, at this moment, staring back at them?

I return to bed, one thought after another competing for my attention, until I drift off.

Chapter Sixteen: The Cat that Roared!

We convene early Thursday morning and discuss our options. The next scientific venue we are considering for release of the information is in Woodlands, Texas. It is billed as the forty-sixth "Lunar and Planetary Conference" and will be held in Woodlands, Texas at the Woodlands Waterway Marriot Hotel and Convention Center from March sixteenth through the twentieth. The last day of the conference will be on Friday.

"We could make the announcement anywhere, anytime we want," says Linda. "Anyone would put their agenda aside to accommodate us, with the news we have. We could even call a press conference today if we want."

"Does it matter where we announce our findings?" I say. "Wherever we decide, someone is bound to feel slighted. What do you think?"

"I think we owe it to Syed," says Linda, "to make the announcement here in Cairo. They are 'Egyptian' tablets after all."

I agree with Linda, Syed should have "first dibs." "Would you like that Syed? Would that be alright?"

"Yes, that would be splendid. We will need to decide on an appropriate forum. Cairo University had previously sponsored a conference for scientific research. We had very good participation. We welcomed many researchers from Egypt and the Arab States as well as distinguished visitors from Europe and America."

"How do we make it happen?" I ask, already knowing the answer.

"I will need to make some calls," says Syed. "Discreetly of course."

Linda looks at me and asks the question on both our minds: "There is still the matter of notifying the university and seeing what their reaction is going to be. I don't know. Should we wait, postpone everything until everyone has an opportunity to participate?"

Now that we have decided on a course of action, we are anxious to see it through. "That could take a good deal of time," says Syed. "We would need to coordinate our efforts and schedule much in advance."

"We are all interested in the same thing," says Linda, "and I am as eager as anyone to get this information released, but we may want to consider how this will affect the scientific community, not just the universities. Perhaps we should wait until we can organize everything more carefully."

Deferring once again to Syed, I ask him for his thoughts on postponing the announcement until we can notify our scientific colleagues and allow them to attend.

"That would be better," he says. "We must not give too much away, however. We will need to proceed cautiously."

We agree it would be best to coordinate our efforts and plan a more organized presentation. It makes sense. In our enthusiasm, we have run ahead of ourselves.

Syed sets up a press conference in the *Grand Celebration Hall* at the Cairo University for Wednesday evening at seven p.m. We figure the information is of such magnitude that anyone who wants to attend will rearrange their schedule to be there. In the meantime, I notify the President of Stanford about what we are doing and he says he will honor my request for discretion. He is beside himself with the excitement of the news and what it might mean for the university. He pushes for more details but I stand my ground. I'm not about to 'spill the beans' before the big event.

"We may be missing someone," says Linda.

"Who is that?" asks Syed.

"Oh, just the President of the United States," she says. "He might want to know."

Syed and I have a good laugh at Linda's cavalier delivery.

Syed says he will contact President Abdel Fattah el-Sisi and leave the matter to him.

Syed can pick up the phone and call the President of Egypt? Why not?

Linda looks at me for a moment as if she is going to object to Syed's proposal but then withdraws. This entire project is, we had agreed, Syed's "baby."

"So, what did you say?" asks Linda, inquiring about my call to the university.

"I told the president about our intentions to make an announcement; that we had a major find and I was reluctant to release too many details at this time."

"And he went for it, just like that?"

"He asked me to reconsider the way we are going about it and asked for more details. I told him it wasn't just up to me, that a professor from Cairo—I didn't mention his name and, thankfully, he didn't ask—was responsible for the discovery and I am following his direction. I told him we had set up the press conference for Wednesday evening at seven p.m. Cairo time. He bought it and we let it go at that."

"Wednesday" Linda mutters. "That should work, I guess."

"What are you thinking about?"

"I probably should be back by then."

It is now my turn to be the "string" puller. "I took care of it," I say, throwing in one of Syed's pat phrases: "Not to worry."

Linda laughs. Syed smiles. I am relieved. I thought he might think I was making fun of him. He didn't.

"I suppose it doesn't really matter if everybody gets wind of what's coming," says Linda. "They don't know anything, any details. It will just add to the excitement when we make the announcement."

"We need to settle on which documents we will highlight during the conference," I say, getting back to the basics. "I prefer the one covering 'dark matter,' for starters. I don't think we should overdo it. All of this information will be released after the conference."

"Maybe we should have just one document for each 'marker' " says Linda. "I would like to present the one on learning acquisition showing how neurological changes associated with learning can be integrated directly, bypassing the actual learning process. I think that would be awesome. It's a bit involved but I think I can pull it off."

"That takes care of two of the 'markers.' Any ideas on the third?" I ask.

"That would be your area, Jake," says Syed.

"But I already have a talk on dark matter."

I can tell Syed isn't crazy about presenting any of our information at the conference. He couldn't be shy, could he? It doesn't matter. "I'll think of something," I say. We leave it at that.

What is information? I know what I think it means—knowledge obtained after struggling with concepts and ideas, hypotheses that admit to a resolution. Something *new*, something never before imagined, ownership of which allows us to say, "I didn't know that before and now I do." Information is more than a simple transfer of facts from one person to another. To an engineer, information means one thing. To a computer software developer, it means something altogether different; the former uses the information to construct; the latter to instruct. But information is more than data; information is built into the very fabric of the universe. It speaks to us every waking moment; sometimes nudging us to appreciate what we thought we already knew in a different light, to see things differently, no longer taking for granted what has become commonplace in our lives. It is our challenge, our opportunity, our obligation to wake up from our stupor and take a leap of faith and see what we have never dared to see before. It is then, and only then, that we truly come to understand that everything—looked at differently, in a new light—is information, everything is new, waiting for us to see it for the first time.

Having taken it upon himself to organize the conference, Syed spends most of the day at Cairo University. He left instructions for Emile to take us into the city.

Emile is friendlier than usual. I don't know if Syed had spoken to him or his attitude toward us has changed for some other reason. Either way, he is much more pleasant and helpful than on our last excursion when he took us to the Club Absolute. We ask him where he would recommend we go for some pleasant "walking around" time in Cairo and he suggests we visit the Khan el-Khalili bazaar. He takes us there and we walk some way until we reach the market. I tell Emile we will call when we are ready to head back to Syed's home.

The market is overflowing with Egyptian souvenirs, tangled alleyways, and vendors, lots of vendors. We are barraged by sellers of wares hawking spices, jewelry, and trinkets galore. Showing interest is a signal for vendors to pounce. We quickly learn to walk away rather than engage the vendors when they become overly eager to make a sale. We find the overall experience enjoyable however and afterward we stop at the Fishawai's Café and sample a cup of tea.

"It's a good thing we had coffee before we left," says Linda. "I would be having a coffee headache right about now."

We are both hungry and ask for a menu.

"This dish looks okay," says Linda. "It's called 'Fattah.' Isn't that the president's name?"

"Yeah, I think so. Maybe his parents loved the dish and named him after it."

"I don't think it works like that. They're not as fickle as Westerners."

We order "Fattah" which comes with "Baladi" bread. The dish contains meat and rice with some type of vinegar. It is palatable.

"Are you looking forward to Wednesday?" I ask. "Got everything nailed down for your presentation?"

"Not yet but I'll get there. The document is very interesting."

"We didn't finish reviewing everything," I say. "Do you think there might be something else in the documents we've overlooked that might be important?"

Well, it's just . . . no, not just . . . I'm sure more documents will show us even more discoveries but we can—"

"No, I mean something that we should know now *before* we release the documents."

"You mean like a warning?"

"Something like that . . . I don't know; I just have a feeling we're missing something."

"That's just your paranoia coming out," says Linda.

"It has served me well so far."

We laugh briefly and then pause for a moment, looking at each other.

"Nah," we say, in sync.

We call Emile and arrive back at Syed's home by late afternoon. Syed is on the phone when we arrive and we are both tired. I give Syed the universal fake yawn and stretched arms gesture. He nods to let me know he understands. We go to our rooms and I take my customary power nap.

After a brief nap, I come into the living room to find Syed and Linda discussing the conference arrangements. I ask Syed if everything is all set for the conference and he goes over the details a second time for my benefit. After Syed lays out the venue and goes over the protocols, he mentions that it would be a good idea to have a mock run-through before the actual presentation. Although there is no prohibition against Muslims working on Friday, the Koran clearly states that a Muslim must "leave off business" after the call to prayer, so we settle on Saturday afternoon for our rehearsal.

I ask Syed how much the press and university know about the details of the find. He said he had not mentioned any specific documents or given any information about how we came into possession of the tablets. The university is willing to accommodate Syed's request based solely on his word.

Linda and I think it would be best if we retreated from our sightseeing and "tourist" mode and concentrated on matters at hand. We spend the remainder of the day and Friday polishing our presentations and gearing up for Saturday's rehearsal. By the time Saturday afternoon comes around, we feel confident we are prepared.

The first time Linda sees the Cairo University up close she is as impressed as I had been. Syed takes us to the *Grand Celebration Hall* and shows us how the equipment works and walks us through the timeline and procedures we will need to follow during the presentation. He then excuses himself, saying he has a short errand to attend to but will join us shortly. As Linda and I look around the auditorium, picturing a capacity crowd, the weight of responsibility on our shoulders to get this right begins to sink in.

Linda walks over to the dais and begins to set up in preparation for the practice run. After taking a moment to collect her composure she places the drive in the laptop computer. She then looks at the computer screen in front of her and sees a cascade of images, all blank. Each image is framed with a dark rectangular outline, inside of which, nothing—no information whatsoever! "There's something wrong with the equipment," says Linda. "It doesn't matter. I have my . . . No, I don't have any notes. What was I going to talk about? " Linda rubs her forehead, and then wrings her hands in desperation, looking lost, totally lost. "Where is the information? Where is it?"

It is clear to me: The information is gone.

We look at each other for a moment, recognizing what has happened. We both know the information is gone . . . everything. "Where is Syed?" I ask, hoping he might have an explanation.

"I don't know. I haven't seen him. I've been busy concentrating on my presentation."

I call Syed.

"Jake, is everything alright? How is the rehearsal going?"

"Where are you?"

"I am on my way to the research center."

"We may be missing some information," I say, minimizing what has just happened.

Syed sounds concerned, but calm. "I will be there as soon as I can."

"Did Syed know anything?" asks Linda.

"He should be here shortly. He doesn't know any more than we do."

"Why was he going to the research center?"

"I don't know. I didn't ask."

"The plates are still at his house, aren't they?" Linda is beginning to put the pieces together, searching for a clue, any clue that might explain what had happened.

"As far as I know," I say. "We can ask him when he gets here. Do you think something happened to them? That wouldn't explain anything, would it?"

"Who knows. I'm just trying to think. Maybe it doesn't matter."

Syed comes into the auditorium and stops for a moment before continuing, trying to get his bearings. "Where is the information?" he asks. "Where are the documents?"

"We don't know," I say.

Syed comes over to where Linda had set up the laptop computer and sees what she had seen, a sequence of blank documents. "What happened to the documents?" he says, staring at the screen with a look of incredulity, giving way to alarm. "How could this happen? There must be some mistake. Do you have the right—"

"It's gone, Syed. It's all gone. We don't know what happened. We have the right thumb drive. There's nothing on it." I couldn't think of anything else to do but tell Syed the truth. There is no covering it up even if we wanted to and I have no intention of doing so. It happened. I didn't know why, but I did know we had to deal with what had happened, somehow.

Linda looks at me and Syed, puzzled, stymied; unable to comprehend the enormity of what had happened. "Syed, do you have the tablets? Are they with you? Did you leave them home?"

Linda is practically grilling Syed. Is she accusing him? "Linda, Syed doesn't know any more than we do."

"The tablets are at the research center," says Syed. "I dropped them off earlier and was on my way to pick them up when you called. They have some new equipment they wanted to use. They thought it could tell us something about the discs, how they worked."

"They examined the discs?" My mind is racing, searching, probing. They *examined* the tablets, the discs. They made an *observation*. It doesn't make sense. We've already seen the information. It can't be! "Schrödinger's cat!" I yell, alarmed. "*Schrödinger's cat!*"

"What are you talking about?" says Linda. "What does that mean?"

I'm sure Linda has heard the term before. She wouldn't have the details fresh in mind, however.

"Is that a possible explanation?" asks Syed.

"It's the only one I can think of . . . even though it doesn't make any sense."

"Well what is it?" asks Linda, anxious to hear an explanation, any explanation for what has happened.

"I think we should sit down somewhere. This will take some explanation," I say.

Syed takes us to a conference room down the hall from the auditorium. We step inside and turn on the lights. We each take a seat toward the end of a long table in the center of the room.

Syed and Linda look at me, waiting, silent.

It is up to me to explain the unexplainable. I haven't even made sense of what has happened and I now have the task of telling Syed and Linda what I *think* might have happened. It is the only explanation I can think of. Before I begin, Holmes' adage comes to mind: "Whatever remains . . ."

I take a deep breath and begin to explain, as best I can, the nature of the quantum dilemma we are facing. "Erwin Schrödinger published a paper in 1935 attempting to address the then ongoing clash between the macro and micro worlds in classical and quantum physics. In quantum theory, there is a term called 'superposition.' This term refers to the behavior of sub-atomic particles. An object is considered to exist in all possible states simultaneously until a measurement is made. Once there is an observation or measurement the object takes on a single possibility out of all possible states. Erwin Schrödinger offered an analogy to help us understand the concept of 'superposition.' It consists of a theoretical experiment known as 'Schrödinger's cat.' "

No glassy eyes yet. "We place a cat into an enclosed box. We then add a vial of poison. There is also a radioactive substance in the box which, when any atom of the substance decays, triggers a mechanism breaking the vial and killing the cat. An observer outside the box has no way of knowing whether the cat is dead or alive until the box is opened and an observation is made. Until then, the cat is said to be in a superposition of states, both dead and alive. This is sometimes referred to as the 'observer's paradox.' It is the observation or measurement that determines the outcome—no observation, no outcome. The quantum state of any physical system is described by a wavefunction which, in combination with the system's environment, obeys the Schrödinger equation. When we make an observation or measurement the wavefunction is said to 'collapse.' More technically, the system's psi-function essentially blends the living and dead cat until an observation is made, triggering the wavefunction to 'collapse' into a single eigenstate."

Ignoring the complexities of my "dissertation," Linda states emphatically: "But we didn't open the box."

"No, but someone did."

We all know the answer. "The technicians at the research center," Linda concludes. "But how could we have seen the information in the first place without opening the box? What is the 'box' anyway? The tablet? The disc?"

"That's where it gets a little complicated," I say.

"It is my responsibility," says Syed. "I took the tablets to the research center. I did not—"

Linda isn't interested in blame. "We should go to the research center."

That is obviously our next move. We go to the American Research Center in Garden City. Syed introduces us to the director of research at the center, Dr. Ahmed Yaseen. I guess Dr. Yaseen to be in his mid-fifties, his midriff giving way to gravity, his hairline receding, and his hair a tad darker in some spots than others.

Syed quickly and uncharacteristically dispenses with the pleasantries. "What have you discovered?" he asks, in English.

Dr. Yaseen looks upset at Syed's abrupt manner but thinks better of a confrontation. "We have examined the tablet and have found no information. Our inspection was very complete and we have used very modern equipment. We do not find any indications—"

"Do you have the discs?" asks Syed.

"They remain in the tablets. They are not disturbed in any way. We have examined them very carefully."

Syed continues questioning Dr. Yaseen until we realize he has nothing more to tell us. We take the tablets and leave.

We decide to return to Syed's home. On our way back Syed remembers he had taken copies of the documents and placed them in his desk in the studio. There is hope.

When we arrive home Syed hurries immediately to the studio to retrieve the documents. They are gone. The drawer is empty.

"Now this is just getting spooky," says Linda. "There is *no* way this could happen. It's impossible!"

Except it isn't. The evidence is plain, literally, and indisputable.

"Jake," says Linda, in that unmistakable tone she uses when calling me out. "You know I respect your knowledge of physics. I'm not questioning that. But this 'wave' or 'superposition' or whatever; it can't possibly reach out and *erase* information *outside* the box or container. Well, in this case, the tablet. I guess the tablets are the containers . . ."

"I have to agree," says Syed. "I believe I have understood what you have been saying, but it cannot possibly account for what we are seeing. I know these documents were printed correctly."

"Jake, there has to be another explanation," says Linda. "How could we see the 'cat' inside a box, whether it was alive or not? The documents being the cat I suppose. I don't know; it just doesn't make sense and I don't see how it *can* make sense. We copied the documents onto the thumb drive. Why didn't *that* 'collapse' the wavefunction?"

"I don't believe we ever saw the 'cat,' the documents, in the box. What we must have seen was an image, a shadow—or whatever you want to call it—of the 'superposition.' And somehow, by extension—again, I haven't figured it out yet—the collapse of the wavefunction, the observation by the research team, affected *all* the information contained in the documents on the computer. As strange as it sounds, I believe a surface scan was possible without collapsing the wavefunction but a *deeper* scan somehow triggered a 'measurement.' The discs embedded in the tablets were designed in such a way as to allow a scan of the *superposition* of the document images, leaving the actual data intact. The process that collapsed the wavefunction, the deeper scan, penetrated the *superposition* and caused the wavefunction to collapse. We weren't seeing the 'cat,' in the container. We were merely seeing the superposition, the two states—one alive and one dead—corresponding to the documents with and without information. The dark or shaded borders surrounding the empty documents are simply the 'fuzziness' we encounter bridging the gap from the microscopic to the macroscopic or human sale."

"Are you saying there is some kind of Akashic field that is being affected?" asks Linda.

I don't know what she is talking about and don't want to know. "I don't know what that is Linda. I'm trying to deal with what I know, or at least what I think I know."

"Why would they design the discs that way? What were they trying to do?" Linda's questions are valid and I can't come up with an explanation off the top of my head.

"I don't know," I say. I can't think of anything to add. It doesn't make any more sense to me than it did to Linda.

Syed is listening, doing his best to follow our discussion. "I still don't see how we were able to see into the 'box' as you say and look at the documents. You tried to explain but I am not understanding."

I need to put this as simple and straightforward as I can. "We didn't look into the box; we didn't open it. In the Schrödinger cat analogy, the cat is alive when placed into the box. There's no mystery as to what the cat looks like; anyone involved in the experiment sees the cat before placing it inside. That's what we were seeing—the cat, or documents, before being placed into the box. Or, more accurately, images of the documents. Once an observation was made, not only the documents but the images of the documents were extinguished. The documents on the discs and the images of the documents were tethered and, once the disc was breached, both were lost to us, in every sense of the word."

I then add a disclaimer just in case Syed or Linda takes me too seriously. "I fully understand that what I have just said can be demolished by counter-arguments: 'You can't tie images to the actual documents and entangle the entire system.' My speculation as to what happened does not follow the normal line of reasoning and understanding of quantum systems. But that's the whole point—nothing about what has happened follows a logical path. The discs were embedded in a quantum system created to trigger a self-destruct mechanism should anyone tamper with them at a deeper level than a simple scanning device. They did not want anyone extracting the discs or compromising the discs beyond simple scrutiny of the information."

"Why would they do that?" asks Linda.

Ignoring Linda's question, Syed asks "Is there any way to reverse what has happened? Can we recover the documents?"

I understand immediately where Syed is going with his question. Recovering the information is all that mattered.

"Short of a resurrection—"

"But is it possible?" he asks.

My explanation doesn't mesh. I'm missing something. Is everything entangled? The documents, the images, our neurons? We are only now seriously considering whether more than two particles can be entangled; and yet, if the entire universe has its own wavefunction, I suppose anything is possible.

Linda looks at me tentatively, perhaps wondering if she should ask me what is on her, and possibly Syed's, mind. "Are there other possibilities?" Linda asks, the implication clear.

My feelings are not hurt. "I could call Jeremy and see what he might think but of course that would be letting the 'cat out of the bag.' " I couldn't resist. Neither Syed nor Linda appreciate my warped sense of humor.

Syed doesn't know who I am referring to and doesn't seem to care. "If you believe it could help," he says.

"It's not going to make any difference if he knows," says Linda. "Everybody will know soon enough. He just needs to keep it confidential for a day or two until we can figure out how to handle everything. Can you trust him?"

I don't feel the need to answer Linda's question. I do some mental calculations. "It would be about ten in the morning in California. I'll see if I can reach him."

"Does he know anything about quantum physics?" asks Linda. "I thought he was an astrophysicist."

"Sure," I reply. "It's par for the course. We've had several discussions about 'black hole complementarity' and other problems in quantum physics."

"We have nothing to lose," says Linda, "that we haven't lost already."

Jeremy and I are not close friends. We have a good professional relationship and take lunches together on occasion. We love to pick each other's brains on scientific topics that interest us. Jeremy is in his late forties but could pass for his late thirties. He has boyish good looks and a terrific sense of humor. He reminds me of Tom Cruise in a roundabout way and is about his same height, a couple of inches shorter than me.

"Good point," I say. I call Jeremy and put the phone on speaker.

"Who is this?" says Jeremy.

Who is this? Doesn't he have a caller ID? Well, I am calling from overseas. No matter. "It's me, Jake Banner. I'm in Cairo. I need to pick your brain."

"I'm driving," he says. "But it's okay; go ahead."

"You're sure?"

"It's the twenty-first century Jake. We have the technology now."

I laugh briefly but quickly compose myself. "Listen, I have to run something past you. It's very important and . . . *very* confidential."

"You got it."

I know Jeremy can sometimes be a little direct. It is part of his personality. But I also know I can trust him. He is just being himself. I start to explain what had happened when Jeremy tells me to hold on for a moment while he pulls over off the road. He wants to give me his full attention.

I gave Jeremy the details as best as I can reconstruct them. "So, what do you think?" I say, unable to come up with another way of asking for his help.

"Maybe you're overthinking Jake. Maybe quantum shenanigans aren't even involved. Perhaps the information stored on the discs was destroyed accidentally when they were examined. As far as the documents missing from the thumb drive; maybe they were stolen or erased. I'm not saying that's what happened. But it's easier to believe than what you are telling me."

I thank Jeremy for his input and give him the dialing instructions to call me if he thinks of anything else.

"So, what do you think?" I say.

"It would seem to me that we have another possibility," Syed offers. "Perhaps we are looking at this in the wrong way."

"Let's assume that Jeremy is correct," Linda interjects. "Where does that leave us? The information on the flash drive has been 'stolen' or 'erased.' The copies of the documents that Syed took have been stolen as well. So, up to this point, what he says sounds plausible; especially when we consider the alternative."

"What Jeremy says about the documents and the copies being taken sounds plausible," I agree. "It's no use trying to figure it out. This is it; this is where we are, in this time and place. We have a presentation pending and we're screwed. There's no way to reconstruct the information. We don't have anything to go on." There, I've laid it all out . . . reality. We just have to deal with it. But why such an elaborate joke? What's the point? To humiliate us?

After my brief pity-party, a wave of anger and determination sweeps over me. "No! I don't know what forces have conspired against us, but they're not going to win. We're going to figure this out. We'll be okay. We just have to put our heads together."

My speech, even in the middle of seeming defeat, sparks a round of enthusiasm. Linda and Syed agree we will not give in. If there is a solution, we will find it.

Linda starts brainstorming. "Did the technicians examine all three tablets?"

"It wouldn't matter," I say. "They're all part of one system." I don't see the point in explaining further.

Syed goes to put away the tablets. When he returns, he exclaims: "I have the copies! They were in the compartment where I keep the tablets. I must have put them away."

Linda and I are beside ourselves. What a stroke of luck. "Back-up hard copies saving the day. All this twenty-first-century technology and scientific breakthroughs and all it took were good old-fashioned Xerox copies."

"Let's go through them and check if we have everything," says Linda.

We agree and Syed begins going through the documents. When we come to the end, three hundred and three copies, we find one extra copy. "Look at this," says Syed, as surprised as Linda and I to see the additional copy.

Syed can read English but hands the document to me to navigate it more quickly.

"It mentions the possibility of an asteroid impact, identifying the asteroid. There is some mention of an 'alternate timeline,' a comparison between present and alternative 'realities,' whatever that means. The rest contains language highlighting some of the details contained in the documents but nothing specifically telling us about its origins. No other explanation. That's it."

"That's it?" asks Linda. "There isn't anything else?" Linda looks at Syed. "I guess you were right," she says. "They must not have been able to prevent the impact." Linda then pauses for a moment, muttering something to herself. "What does 'alternate timeline' mean?"

"I don't know," I say. "I understand the concept but without further details, I can't make heads or tails of it. I would feel strange going in front of the world touting our discovery knowing what's coming. We've known about the asteroid, but this confirmation in the cover letter is unsettling. I don't think we have a choice. It would feel out of place, self-serving. Besides, it would be better if we released the information after a more thorough analysis by others in a better position to evaluate the impact."

We are talking past *what* had happened: How were the tablets 'transported' through space and time? It is an interesting question and one I am seriously interested in answering. And why were the tablets sent to this time period and not another? Why Egypt and not elsewhere? And so on. For the present, however, we have other concerns that overrode our interest in understanding how the tablets made it to us . . . intact.

"Should we cancel the conference then?" asks Syed, seeking consensus.

"So much promise," I say. "Everything will come to nothing if we don't do the right thing and act quickly."

Linda looks despondent, tired. "A lot of people are going to be disappointed about the cancellation. What are we going to tell them? We can't tell them about the impact. That kind of information would spread around the world in a heartbeat."

Syed speaks in a reasoned, diplomatic manner: "The situation is very grave and we must take care to do everything properly. If we tell anyone about this information it will have to be in person. Nor do I think we should tell more than one individual about this event. We must decide."

I understand Syed's reasoning perfectly. One head of state, whoever that might be, should orchestrate the international parties to act in concert. "There are procedures," I say. "Recommendations for an international response will have to be set up. It's a matter of how this plays out, who starts the ball rolling."

Linda wrinkles her forehead, deep in thought. "This isn't some stargazer getting excited about spotting a previously unknown comet in our solar neighborhood. This is a verified asteroid with a possibly devastating, even world-ending, impact. I think we should notify . . . No, wait. We have to go through channels. We can't just call up the President. What's the protocol?"

Good question. However critical the situation, we need to handle this the right way. "Maybe we should get hold of the Pentagon in Washington," I suggest. "Would they even take our call? Despite our bona fides, we're not exactly at the top of the food chain. They would want more to go on before they let us talk to the Secretary of Defense. Come to think of it, does NASA even report to the Defense Department?"

"I could call President Abdel Fattah el-Sisi and he could contact your President," offers Syed. "We would then know how to proceed. I cannot think of a better plan."

We all agree. Syed leaves the room and makes the call.

"It's not too late in the afternoon," I say. "Syed should be able to reach President el-Sisi. He was already told about the conference. After the Presidents contact each other, we should have a better idea of what to do about the conference. Maybe they will want us to carry on as usual; you know, so nobody would be speculating about what's going on."

"I guess," says Linda, letting out a deep sigh and tilting her head back on the sofa, closing her eyes for a brief moment. "I'm all wired up and tired at the same time. I need to relax."

After a good while, Syed returns and shares his conversation with President el-Sisi.

Syed tells us the Egyptian President requested a meeting immediately. "I did not mention the specific nature of the threat. I did not know if the line would be secure. But the President is aware that we are facing a grave danger. He is sending a car for us which should be here shortly."

A limousine soon arrives, followed by a military escort vehicle. I guess we aren't maintaining a low profile anymore. The driver opens the door for us and we get in. The limo and escort take off. Syed tells us the destination will most likely be one of the President's main residences, the Heliopolis Palace. He explains that the palace was formerly a hotel until it was converted. The vehicle seems secure enough, and heavy. Probably reinforced.

Although I remain composed as we drive to the palace, I am becoming ever more aware of the scope and magnitude of what is happening. I think of the adage: "Be careful what you wish for." I had wanted excitement, a change from my professorial routine. It has arrived. Now what?

As the driver approaches the palace I take in the site—a synthesis of Persian, Islamic, and neoclassical architecture. The tall archways and the Central Hall's dome are prominent. We enter through a side entrance and are taken directly to the President's office.

What the president lacks in stature he makes up for in charm and charismatic personality. After the President and Syed exchange a warm greeting and a few pleasantries, he turns his attention to us. Speaking through a "trusted" translator, he welcomes us to his "home," one of many, and offers us some coffee. To our surprise, the coffee is typical of what we would expect back home.

Syed explains what we had discovered about the asteroid and when we should expect its arrival. The President's command of international procedures is exceptional. He speaks briefly of the UN recommendations for an international response and goes into great detail concerning an *Interagency Deliberate Planning Exercise After Action Report* from 2008. The exercise, he informs us, was coordinated by the Pentagon at the *Directorate of Strategic Planning,* United States Air Force. A scenario was developed wherein an imaginary asteroid called *Innoculatus* was sited a mere seventy-two hours before impact. Teams were organized to come up with appropriate protocols and procedures for a coordinated response. Nothing concrete came of the conference he says but protocols and procedures are ongoing.

Sounds about right. When an asteroid is about to crash into our planet, then we'll get serious.

Although speaking through a translator, it is clear to us that the President has a keen grasp of the situation and understands protocols. The President asks about the information but Syed gives very little away. Syed does mention the information is not from the present which the President readily grasps. He does not however go into any details about our findings other than to inform the President that the documents are extraordinary and of great scientific value.

Syed then asks about our next move. President el-Sisi tells us he will be speaking with the U.S. President to create an appropriate action plan. He also tells us he will stress the importance of our continued involvement. He believes we deserve to be part of the international effort to seek an appropriate response to the threat. As far as the conference is concerned, he believes it would be best to continue with our plans. He will however discuss the matter with the American President before making a final decision. With that, we pay our respects and the limo takes us back to Syed's home.

There is something about the documents, especially the "cover" letter which strikes me as familiar—very familiar. But I can't quite put my finger on it. I ask Syed for the documents and he brings them to me. I know what it was: The cover letter explaining why the documents were preserved is familiar to me. The writing reminds me of someone; I can't quite place it. Maybe I know the person who drafted the letter; that could be. I look at the letter again holding it up to the light. There is some writing or text of some kind. It is written in binary code in very small print.

01000001
01000010
01000011

Linda is out of the room. I call her and Syed to come over and look at the document with me. "This is simple binary code," I say.

"What does it mean?" asks Syed.

"I don't think it's referring to numbers. That wouldn't mean anything, just numbers based on binary math. Wait a minute. Linda, can you look up the English binary alphabet on the net? I have a hunch."

Linda retrieves the laptop and looks up the information. "It's just the first three letters . . . capital letters, of the alphabet. That could be anybody."

"Yeah I guess you're right," I say. "It probably doesn't mean anything."

"The first three letters of the alphabet," repeats Syed. "Why would anyone put such information on the document? It doesn't explain anything. What is the point of it? It must mean *something*," he insists. "How could anyone know who they were? They might just as well have left it blank, with no authorship. There must be something missing."

"Azad, Banner, Cooper," says Linda. "We're responsible!"

We have a good laugh . . . and then it goes quiet.

My brain is trying to work, nudging me to put two and two together. "The tablets were found somewhere in or around Cairo, or so we believe. Now hold on a minute; let me finish my thought before it goes away. This Khaled fellow, for some reason, thought Syed would be interested in the tablets. It could be a coincidence. I'm not saying it isn't. It's just a factor in the equation. Now . . . let's say we *are* the authors of the warning letter and maybe even had something to do with gathering the data. If we could send them back in time and we wanted someone to know about what we found out, who would it be?"

"It would be us," says Syed. "We would want the information so we could take action. But we would not know of the events before receiving the documents." Syed takes my scenario further but seems stymied, unable to complete his thoughts.

Linda has been following what we were saying perfectly. "We sent the information, the documents, to us—to Syed—so he would have them and we could take action after understanding what happened. Yes, we would be in the dark as to why we had the tablet, but we would, again as we did, figure everything out. Maybe they were specifically sent to us but—"

"They were stolen!" I say. "We only recovered them by happenstance. They could have been lost forever."

"So maybe Khaled knows more than he let on," says Linda. "But if he and his gang stole the tablets from Syed, Syed would have known about it. And Khaled would know he was taking a risk in trying to sell them back to Syed." Linda pauses for a moment, shaking her head as if she is trying to dislodge whatever logic she has remaining. "Maybe he didn't steal them. One of his accomplices could have. They could have been taken, or stolen *before* Syed even knew he had them . . . No . . . I don't know."

"That could be right," I say. "It makes sense. But it wouldn't matter one way or the other if Khaled knew that Syed knew and so on. Khaled was willing to take the risk for the money. He probably didn't care if Syed believed they stole them or not."

"It sounds like that may be so," says Syed. "We may not find out for certain. We are fortunate to have the information we do. We must now concentrate on what we have left to do."

Something just struck me. "Linda, how come your last name is still Cooper? You never got married?" As usual, a moment after my comment, its chauvinistic nature becomes obvious.

"Maybe I kept my maiden name for personal or professional reasons," Linda says.

"Sorry, I wasn't thinking." Note to self: We are in the twenty-first century, Jake.

"I know the horse is already dead," says Linda, Syed looking on puzzled but choosing not to interrupt for an explanation. "But, if the documents were 'sent' directly to Syed's home, there would be some record of a break-in, something on the surveillance camera. The break-in would have been random, a coincidence. Khaled, or whomever, wouldn't have known about the tablets beforehand. And how could anyone get past the security—"

"I don't know how we're ever going to figure out everything that happened," I say. "Maybe they weren't sent directly to Syed's home but somewhere else where he might find them. I'm with Syed—we're lucky, we have the information and we need to stay focused on the issues at hand." Syed and Linda nod in agreement.

We have no idea, of course, who collated the documents and 'sent' them to us in Cairo. Was it just our egos making assumptions? It does make some sense however and cannot be dismissed lightly. Syed would be in his mid-eighties by the time these future events would take place. By then, science and medicine—as we have seen in the document cache—would have advanced to the stage that someone in their mid-eighties might still be relatively healthy and productive. My brain starts to tie up in knots. I decide to redirect my attention and energy toward what I hope will be a more productive use of my talents. My "mental shelf" is getting crowded.

We come to a consensus: Stay the course, complete our plans for the conference, and wait for further word from President el-Sisi.

Chapter Seventeen: Paradox

By early Sunday afternoon, we are ready to take another run-through for our rehearsal. We had scanned and uploaded the copies and made additional back up discs. We are about to head out to Cairo University when Syed receives a call. It is from an assistant to President el-Sisi. He explains that the President is requesting we cancel our plans for the conference. He says the President will send a car for us shortly.

"What could this be about?" asks Linda. "He gave us the impression that he wanted us to go ahead with our plans."

Syed reminds us that President el-Sisi wants to discuss the matter with the U. S. president before giving the final go-ahead and has not yet done so.

"So there has been a change in plans," I say. "I guess we will soon know why." I have a suspicion but it would probably be better to wait and see what President el-Sisi has to say.

When we arrive back at the President's palace we are told, again through his translator, that he will handle everything and we need not worry about any press fallout from canceling the conference. He then explains the reason for his request. Much is lost in translation but the gist of his explanation is clear enough—grave concern over a "time paradox."

I had anticipated such concerns but now, hearing them out in the open, it seems obvious. I should have been the one to bring it up. It is embarrassing to think President el-Sisi's scientific advisors had come up with the idea while I had said nothing. Perhaps my thinking has been wrong. It seems to me that, having taken the action we did in the future, we must have worked through such complications already. In any event, we had *already* tampered with the timeline by introducing the tablets from the future into the past, our present. It isn't immediately clear how our actions from this point forward would do anything more than *add* to any paradox *already* breached. But it is worth considering, working through the theories to see where they might lead. Unleashing advanced scientific knowledge prematurely—before its "scheduled" time—could aggravate an already tenuous future.

"It's pretty clear that we have a major find," says Linda. "Our announcement says as much. It's the details that remain secret, so far. Everybody also knows by now that Egyptian tablets are part of the find."

"You're right," I agree. "Although the announcement was cryptic regarding the information content, everyone surely knows there are tablets involved. Emile knows for sure and the lab technicians at the research center. But nobody has any specific information about the documents. We haven't shown them to anyone."

"This is true," says Syed. "President el-Sisi, and now I am sure your president as well, are aware that we have major scientific findings. But yes, the information they have is not specific."

"Has the U. S. president been notified yet?" I ask Syed.

"I am only hearing from the Egyptian government on this matter. I do not know if the U. S. president has been informed."

That seemed premature. What if the American government had a different idea about what we should do? Would our scientific advisors come up with the same conclusions about a "time paradox"? It seems to be a legitimate concern but surely a different perspective would help. I look at Linda and ask her if she thinks we are missing something, some angle we haven't considered. I immediately regret asking the question.

Linda gets to the heart of the matter: "How important is this information to the superpowers and how anxious would they be to get their hands on it if they knew it existed? And how important is it to the United States and Egypt to keep this matter from being disclosed? Do they trust us? Would you?"

Syed is closer to government intrigue than either of us and I am eager to get his take as well. "What Syed?"

"They must believe this information would be dangerous if it should be released. That seems certain. No, I do not think they would trust us to remain silent. It is too much risk; they would want to have complete control, assurance that everything is under their direction. I do not feel good about our situation."

Linda again gets quickly to the point: "So are you saying we may be in danger? That our governments may not trust us to keep this quiet."

"That is a very good possibility," says Syed.

The old aphorism about paranoia comes to mind. I suddenly become very angry. "So, this is the upshot of our finding a treasure trove of scientific wonders! We have become the targets of our governments."

Linda has been sitting relatively calmly, but seeing me go off is unsettling. "We don't know anything Jake. It's all speculation at this point. We're jumping to conclusions."

I feel like telling Linda she's the one that brought it up but I hold it in. Channeling Dylan, I say: " 'You don't need a weatherman to know which way the wind blows.' " I can always count on my paranoia to see me through.

Syed is concerned but, unlike me, is holding it together. Why couldn't he lose it, just once? He's human too. Isn't he?

In a calming voice, Syed goes into diplomatic speak. "We do not know where this is going. We have reason to be concerned. This is normal, human nature. But we must be . . . level-headed. It would not be wise to, how do you say, 'jump to conclusions.' "

I wish I had Syed's composure, but I don't. I don't trust the government, any government. They have been set up to govern, to control society. That's why they have a monopoly on power and also why they will do most—no make that *anything*—to remain in power, to remain in control. It is Hobbes's Leviathan writ large. I have been trying to keep my paranoia in check but am beginning to realize I needn't have.

"What do we do if they ask for the information?" asks Linda.

Syed and I look at each other hoping the other will answer. My mind answers without thinking. "It's not the government's information, it's ours. We have control over what *we* do with it."

I have never seen Linda look at me the way she is now. Newfound admiration? I can only hope. I didn't think I had it in me. Maybe I am a leader! Maybe I am heroic! Nah, on second thought, I should just stick with paranoia and forget the delusions of grandeur. Although I must admit, it feels pretty good.

"That is a very good point," says Syed. "It might also present an opportunity to test our suspicions. Let us see what they do. If they demand the documents and we refuse, we will see for ourselves what their intentions are."

"That could be risky," says Linda, looking to Syed for a response. "Don't you think?"

"Yes."

It always makes me nervous when Syed gives a one-word answer. A curious thought enters my mind: "What does any of this even mean?" I ask. "They can't 'do us in.' We wouldn't be there in the future to make any of this happen." I start to get flashes of my undergrad days learning Fortran, the scientific computer language. Nested loops within loops, within loops . . .

"What are they so worried about?" says Linda. "What is this 'paradox' supposed to do, destroy civilization? If what we think is true about what happens in the future, we're all pretty much done for anyway, right? Isn't that the whole reason we—well, maybe us—gave the warning in the first place? What do they think would happen if people knew about the documents?"

Linda's mind is the "gift that keeps on giving." She's right of course. What is the problem they are so worried about? I have to give this some serious thought. I must be missing something. I know about time paradoxes; there are more than one. What, specifically, would be worse than . . . ? "I need to do some more research on this."

"We could just ask them," says Linda, "what they are concerned about. Why it's so important. Wouldn't they tell us?"

My mind begins to sift through the implications: Why didn't they give us more specifics in the first place? The way they had put it, there was simply a problem with releasing the information due to a "paradox." They didn't go into details and—with Syed as the go-between—I hadn't pressed them about it. Maybe I should have. Their concerns could be legitimate, who knows.

Syed continues his diplomatic assessment. "I should be clear," he says. "there is no good reason to suspect our presidents. They would have the interests of the people to consider. I am sure they would want to do the right thing to protect us. I cannot say with such certainty about those who might advise them, however. There is no way to be sure of what they may think. Sometimes they do not think so clearly. If we are of one opinion as to what to do about the information, they may agree with our assessment. If they do not agree, it does not mean we are right and they are wrong. They may know more than we do about such things."

Well, that's just great! Now we're stuck squarely in the middle. Speaking of paradoxes, I am beginning to feel like Buridan's ass, trapped "betwixt and between." I excuse myself and go into the study to look up the latest research on time paradoxes. Linda comes and gets me after a few minutes. "Syed is on the phone with the president's office," she says.

"Could you make anything out?"

"No, it's all in Arabic. We'll have to wait and see what's going on."

We can't tell what Syed is saying but he is certainly animated. His voice pitches up and down as he makes some point or other. I can't tell if he is winning or losing if that is even the right way to think about it. After he completes the call he sits down and tells us what was discussed.

"As we suspected," he begins, "the administration wants to send security to pick up the documents. I spoke with Mr. Halabi and he said it was—I think the translation would be 'imperative'—that we provide the information immediately. I told him we would need some time to complete our analysis. We would then consider the request. He was not happy to hear what I had to say. I think he will consider what to do next, and so must we."

Time paradoxes are not my thing. I am more comfortable using Einstein's equations to find the Einstein tensor, derived from the stress-energy tensor, throwing in the Riemannian curvature tensor which encodes information relative to the curvature of spacetime from one point to another and then using the components to construct "Christoffel" connections, and so on, eventually determining the direct route objects take through space. A difficult process to learn but, once learned, it comes naturally enough. But this time paradox business; well, let's just say it needs some work. I wish Linda had already gone through this learning curve. It would have saved us some time. Linda and I are both what you might call autodidacts. She usually learns quicker than me, but once I get it, I get it. You could ask me some esoteric question about a process no one cared about or studied and, if I had studied it even years ago, it would still be fresh in mind.

I spend the next few hours boning up on time paradoxes. By early afternoon I am depleted and decide to take my power nap. As I am about to head for slumber my phone rings. It is the dean of my department back at Stanford. She is not a happy camper. She had been told about the cancellation of the press conference and I do my best to explain that it was out of my hands. Although she understands, begrudgingly, she lets me know in no uncertain terms how disappointed she is. I did my best to make her see things from my perspective but I don't think I succeeded. Well, at least my tenure isn't in jeopardy; although I still need my brain.

After my much-needed nap, I wake up thinking about Khaled and his cronies. I wonder what their take is on all of this. Did they get wind of the importance of the tablets? Something else to worry about.

When I come into the main room it is empty. Syed had taken to the study to do some work. I join him and ask about Khaled and the possibility that he and his buddies might have figured out how important the tablets are. He says they're too dumb to bother worrying about. That is good enough for me but I still reserve the right to worry about it.

Linda isn't around so I go and knock on her door. I want to share some preliminary thoughts with her and Syed about what I had put together so far on this paradox thing. She is glad to hear my take on this whole matter. I ask Syed to join us in the main room and we sit down. Syed and Linda are all ears.

I decide to begin on a lighter note: "I think this is where my being a bit 'loopy' might come in handy," I say. "It isn't clear from our discussion exactly which time paradox they are concerned about. There are several versions and many twists and turns these paradoxes can take. One is known as the 'predestination paradox.' It's also known as a 'causal loop.' "

"Oh, I get it now," says Linda. Syed either doesn't know or care what that was all about, remaining silent waiting for something more substantial.

"It has to do with the sequence of events. Ordinarily, one event is followed by another—A causes B." A good start as far as I am concerned, simple, and easy to follow. Maybe I should just summarize; forget the details. We'll just get lost in the woods. "Forget all that, what I just said. Here's the bottom line: History doesn't change. If something *happened* it *must* have happened. Whether we intend to alter the past or not, we can't change anything. The tablets are already in the past which means they *were* in the past before. Their being here is critical to the future and these tablets somehow cause the future to happen the same way we remember it. I know this sounds strange but the crux of the matter is: What happened is going to happen, period. There's nothing we can do to change it."

"Then what difference does it make if the information, the documents, are opened up to the world?" says Linda.

"Ah, but that's just it," I say. "It's a *theory*. Not like the 'theory of relativity,' which has been proven to most people's satisfaction, but more like a hypothesis or idea of what physicists believe will or could happen. They're probably thinking, 'Why take chances?' "

Linda catches the contradiction instantly. "So, they think, *contrary* to the theory, we might very well be tampering with the future by releasing the documents? I don't think they know what they are doing."

"You could be on to something," I say sarcastically. "But as I said, we don't know which specific concern they have. You're right about the 'predestination paradox.' It wouldn't make much sense to try and thwart events if that particular theory is correct. At least we're not in a 'temporal causality loop.' "

"You mean like *Groundhog Day*?" asks Linda.

"I have not seen this movie," says Syed. "Is it very good?"

Linda briefly describes the movie to Syed.

Syed takes the description seriously. "We may have had this discussion before, you think? Is this possible?"

"I hadn't taken that possibility into account," I say. "I hope not. I guess there's no way of really knowing."

Linda has had enough serious talk for one day. "I think that movie was about Zen Buddhism," she says. "He had to keep living the same day until he 'got it right.' It was only when he ended up being a 'mensch' that he was able to advance to the next day."

"What is a 'mensch'?" asks Syed.

"I think it's a Yiddish term meaning a person of integrity," says Linda.

"It sounds like a good movie. I will have to look for it; add it to my collection."

"Don't get him started," I say.

"What other kinds of paradoxes might there be?" asks Syed.

"There are other types of 'temporal loops' where events are repeated," I say.

"Like in that *Star Trek* episode," says Linda. "Where the *Enterprise* and another ship kept exploding. But they escaped; they broke the cycle."

"That is a possibility I guess; at least in the movies and TV."

"So, what else are they worried about?" asks Linda.

"There's the standard 'grandfather paradox.' You know, a time traveler goes back in time and kills his grandfather before he meets his grandmother so he's never born. There are ways around the paradox. One simply states that the past can't be changed so the grandfather must have survived the attempt on his life. Or the time traveler creates another timeline where he was never born. This is where parallel universes come in."

"Where the universe splits and we have multiple worlds and futures," says Linda. "I've always wondered where all that matter and energy come from to—"

"We're not adding anything," I say. "We're just realizing potentials." As much as I don't want to argue with Linda about her comment, the statement that we are scooping up new matter and energy to "create" new worlds irks me. I better get back on track before we do get in an argument. "Besides the 'Grandfather Paradox,' there are several other problems that crop up regardless of what paradox we're talking about."

"If we were to go back in time," says Syed, "we would be seeing ourselves would we not?"

"That is one of the objections," I say. "Another has to do with the present moment. If this moment is all there is, the objection goes, how do we get to another moment in time?"

Syed isn't fooled by any of this mumbo-jumbo. "What are you thinking Jake? It must be possible since we have the evidence before us."

I could have gone into even more paradoxes but this avenue of inquiry doesn't seem to be getting us anywhere. "As 'Ritty' said we may have to live with not knowing."

"Who is that?" asks Syed.

"It doesn't matter Syed. We will just have to deal with the situation the way it is."

Syed is not quite ready to move on. "So, we are saying we have two timelines: one where we do not have the tablets and one where we do. In one future we do not send the tablets back in time to us and in another we do. But would that not mean that the tablets were here before they were sent? There is certainly no reason in the future for us to do anything. It has already been done. Our only concern presently is to see what we can do about this asteroid and somehow prevent it from hitting the Earth. The future has changed the past and now we must decide how we will change the future but only in a limited way."

I might have put it differently and there is certainly more I could add about timelines and altering events but why bother, Syed has captured the essence.

"Back to reality," says Linda. "I think whatever is going to happen is going to happen. We can't predetermine the future any more than we can change the past. If this paradox theory is correct, the tablets are supposed to be here in this time and place. And whatever we do or do not do with them has, in some weird sense, already happened. The future is over; we're just experiencing a rerun."

Syed and I look at each other and burst out together in laughter. Linda joins in.

"It is like I have already told you, Jake," says Syed. "You should retire. Everything has already happened. You have already lived your life and died. I was sorry to miss your funeral but I had died before you so you will forgive me for not being there. I am sure there were a great many in attendance. It must have been very sad."

More laughter.

I have known a few Muslims on a casual or professional basis. Is Syed different? Or had I invented my impressions based on nothing more than sensationalized media reports? Is it simply a matter of the personal distance we create to separate ourselves from "foreigners;" from the "Other"? Is it really so strange to learn that Muslims have a sense of humor; that they feel pain and pleasure; that they laugh and cry and worry over the same things we do? Why shouldn't they? We're all just people trying to do the best we can to negotiate the life cycle. A little humor roughs out the sharper edges. What could be wrong with that?

We talk for a while longer about what we should do with the documents until finally "deciding" to give them up. Whatever we do we were supposed to do. How could it be otherwise?

Chapter Eighteen: Inevitable

I again wake in the middle of the night, restless, uneasy. Something keeps nagging at me, pulling me—something unresolved. I go downstairs to the studio and fire up the computer. I insert one of the discs and bring up the documents relating to the genetic 'marker.' I navigate immediately to document number eighty-nine. I have no idea why I am compelled to look at this document and no other.

The document outlines a diagram depicting the flow of information in the brain—what happens neurologically—when we make decisions. The information is crystal clear: Our "decisions" are driven automatically, beneath the surface of consciousness, beyond our awareness. The information harkened back to Libet's experiments but took the conclusion much further, outlining the specific neurological substrate or underpinnings of conscious "after-the-fact" choices in which we engage without conscious volition. But there is more, much more. The entire epigenetic, neurological, and biochemical processes—including environmental and social factors—that trigger and are responsible for every action we take are documented in great detail. Our consciousness, an "accident" of evolutionary development, is nothing more than a depository of prior subconscious "decisions" formulated beyond the reach of real-time awareness; beyond even the possibility of intervention. Our consciousness serves as a reflective program for evaluating previous mental actions; not as a director or orchestrator of events.

I am familiar with the studies on cognition and action potentials conducted by Benjamin Libet. This work is more complete: It formalizes, documents and confirms the entire mental processes from start to finish, with no room for conscious choice. We "make" decisions subconsciously, are subsequently notified of the decisions we have made, and are then left with but one choice after-the-fact: To take ownership and fabricate, as best possible, the "reasons" for our actions. The report is reassuring in but one respect: Most of our actions or decisions, most of the time, are in agreement with our normal attributes of personality and temperament. However, those selfsame attributes come pre-packaged, predetermined by a plethora of inputs far beyond our control. Our morality or lack thereof comes prepaid.

The report goes on to outline procedures for changing the judicial system to accommodate this new finding. Incarceration would not be eliminated as there would be ample reasons to offload bad actors to institutions for care and treatment. The report also outlines procedures dealing with other issues as well; all tied to our new understanding of how our brains, our neurology works. There are no indications that appropriate and necessary actions to deal effectively with unlawful actors or mental illness would be abrogated in any way due to these new findings. Many Improved and effective measures are engineered to incorporate these new insights and the newer approaches offer hope.

At first, I feel let down at reading these new findings. There seems scant room left for me, for my freedom, my volition. But, upon reflection—that's where I get a chance to play—I feel more liberated, more thankful that my "lot" in life had turned out so well, especially compared to so many others less fortunate. I might not be able to take complete credit when I do the right thing but neither should I berate myself too severely when I fail to do so. I don't look at this information as excusing my behavior when I behave badly but neither do I take it to mean I am an inherently "bad" person if I don't live up to my ideal image of who I think I should be. What I am, what any of us are—what we do or do not do—is, in some deeper sense . . . inevitable. I delete the document on the disc and the backup copies. I couldn't help it. I had to. I had no choice.

We spend so much time wringing our hands, anxious, worried, and frustrated. I call it the "six-months-syndrome": Six months from now when the kids get out of school; when I get the raise my boss promised; "when my ship comes in," I'll be all set. And then, after six months pass, a new horizon appears . . . and another . . . and still another. All the while our lives are being consumed, our minds tangled up in the future or the past; never appreciating the present. That's all we ever really have, this moment . . . now. We can't cause the future before it happens. Were we to exhaust our mental powers we will never "solve" tomorrow's problems today.

My subconscious finally sends me to bed.

I sleep in late Monday morning. By the time I get up Syed is in the living room. He had set the three tablets in a row on the large table in front of the sofas. I come over just as Linda joins us. We gather around the table, Linda and I standing on one side and Syed on the other.

Syed looks somber, reflective. "The president is sending a car to pick up the tablets and the discs," he says. "He has invited us to dinner this evening to thank us personally."

We sit for a while discussing our adventure. Linda and I thank Syed for his hospitality. A car soon arrives and two gentlemen come in to collect the tablets and discs. They also retrieve the copies Syed had made as well. They then thank us and leave.

Linda and I had previously made arrangements for our return trip back to the states. Our flights are scheduled for tomorrow morning around eight a.m. We make sure everything is ready for our trip, complete our packing, and take care of last-minute details.

The dinner that evening with President el-Sisi is cordial, polite, and staid. We have very little interaction with the president due to the language barrier. He does his best to make us feel as comfortable as possible under the circumstances.

It will be our last night in Cairo. I feel a mixture of melancholy and relief—sadness that my adventure is coming to a close and relief that I will soon be back home again.

Linda and I are up early and prepare for departure. Syed, as usual, is up early as well. Emile comes by shortly and we head out. On the way to the airport, our conversation is stilted, awkward. We avoid any mention of the obvious. The normal sad goodbyes await us, especially me, as we drive to the airport.

Once we arrive, we check our luggage. Syed walks with us for a short distance and then stops. He does not wish to follow. Syed and I look at each other for a moment; he shakes my hand and holds on for a moment, then gives me a hug.

"As-Salaam-Alaikum," he says.

"Wa-Alaikum-Salaam," I reply. Syed turns around and walks away. He doesn't look back.

I don't know if what I said was correct or out of line; but if it was, Syed didn't let on. It is the only saying I had memorized in case Syed should express such a sentiment.

Linda and I complete the check-in process and head toward the gate for departure. I have trouble thinking straight, one thought after another tugging at me. Had we made the right decisions? Does it matter? Could it have been otherwise?

Linda is looking forward to getting home again, back to the old routine. We obviously feel very different about what the experience has meant to each of us.

The boarding call finally comes for our return flight via Frankfurt. We board and take our seats. It is surreal. It almost seems to me as if we had done nothing at all. Here we are returning home again . . . with what? What we had uncovered, as fantastic as it was, had not been released; not really. It is in the hands of the Egyptian government. Would it ever truly be shared with the world? And what of the implications for time, information, and the future? Is my assessment correct or is it just speculation? I can't point to anything I know to be correct about how the psi-function operates. Could we see a superposition without somehow tampering with the state of the system? Why were such extraordinary future discoveries embedded in Egyptian tablets? These questions crowd out my ability to think coherently. I am anxious to get Linda's assessment of what has happened.

After the plane taxies the runway and we lift off I sit up straight for a moment, take a deep breath, and try to relax. I want to tell Linda about document number eighty-nine but I'm not sure if I should or not. The whole notion of my being unsure feels strange. If everything is inevitable, then I either will or will not tell her about the document. It's that simple.

"Linda, I need to borrow your brain for a minute."

"Will you give it back when you're through?" she says. We laugh. A good start, I guess.

"I found a document. Not just any document. It wasn't about a medical breakthrough or anything like that. It has to do with the ultimate nature of reality, how our minds process information and make decisions. It was much more advanced than anything I had ever seen. The conclusion tells us that what happens in our lives, what we think we are doing to bring about the results or outcomes we wish, are 'prearranged' for lack of a better word."

"Are you talking about predestination?" Linda looks concerned.

"Not exactly," I say. I then go on to explain the scope of the findings, how we have a limited window to reverse course, and how, upon reflection, we could with some success adjust our course—all under the rubric of predetermined operators. My explanation of the document is nuanced, succinct. I am careful not to exaggerate or diminish the implications.

"Then we have nothing to be concerned about," says Linda. "If the document is correct what we have done and what we will do is 'in the cards.' We are operating with a 'stacked deck.' "

"Of course. I can see how that makes sense. It explains everything: My 'correct' or 'incorrect' understanding of the physics—the psi-function and so on, everything. It doesn't matter." An outlier thought then introduces itself: "If we believe we 'sent' the documents to us from the future then our future selves had in mind what we should do with this information."

This is my opportunity to open the "window" where I can think about what had happened—preordained or otherwise—and put things in perspective. I have to unravel my thoughts quickly before the thread is gone. "Don't you see," I say. "What we, in this timeframe, do may or may not coincide with what our future selves intend. We are younger versions of us. What we think we should do with this information, release it or not—make it public—may not sync with what we will eventually wish we would have done." I can't put it any clearer. It is muddled and Linda will just have to fill in the gaps and understand what I mean.

Linda stares at me with a serious look which tells me she does understand. "So, we have to think the way we would in the future, being older and more mature; more experienced and knowledgeable about what is happening. But why should it matter? You just said—"

"No, we do have a chance to understand, to do the right thing, to make the right decision. Whether it is determined or not, at this moment we are 'outside' of predetermined causality. We are in the space, the gap that allows us to think—to reason. Our decisions turn out to be inevitable if and only if we allow the autopilot, our autonomic program, to run without conscious reflection. In the final analysis, the outcome is driven by our conscious reflection which injects itself into the process. Whether or not we use our conscious reflection to decide is not predetermined. There is a loophole in the machinery of our brains; it is our ability—even if predetermined to an extent—to reflect, to take control of the mind, and to wrestle against our programming. Yes, I am aware of the tautology. It does not escape me: Even when we 'step outside' our programing to orchestrate events we are doing so within the confines of our habituation, or own internal dynamic constructed from and even before birth. But, and this is the important point: The very act of reflecting, of reasoning, is creative and unknown even to us until we engage the process and resolve our conflicts. It is, in this sense, free-floating, unbidden, and unforced. In short, it is un-programed. We still have a chance to think this through and resolve to act as we should, not as we have been programmed to do."

Linda's mind hasn't skipped a synapse. She gets it perfectly. "Okay," she says. "Let's look at this logically. We made our decision based on who we are now and, even though we did wrestle with it somewhat, we didn't take into account what we are now considering: our future selves versus us now. 'We' sent us the information—all of the information—for one purpose only: to release all of it. What other reason could we have had?"

"And we didn't release it," I say. I gently pinch my upper lip with the thumb and forefinger of my left hand, my right hand resting against my side under my left elbow, distilling the information until the only conclusion possible hits me: "We made the wrong decision."

"And now it's too late," says Linda.

"Maybe not."

Linda leans toward me, her left elbow taking the armrest. "What are you talking about?" she says. "It's all gone. We handed it over already."

"I make an extra disc."

"Of course you did," says Linda. "How could it be otherwise?"

We laugh, relieved, and grateful. We still have a chance to make this right.

"It's like having a battle of wits with ourselves," I say. "It's us versus them, and they're both us! It reminds me of the movie The Princess Bride where Wesley, or 'The Man in Black,' is engaged in a battle of wits with Vizzini over which cup contains the 'iocane' poison."

" 'Truly, you have a dizzying intellect,' " says Linda.

' "Wait till I get going!' " Linda is a dream. She likes the same movies I do. She has a great sense of humor. And, as if that isn't enough, she has a great mind and intellect. If she has flaws, and I know she must, I don't want to find out about them until it is too late.

We both welcome the diversion. "There are some movies that should be mandatory, don't you think?" I might as well start.

Joining in the game Linda asks what three movies I would include on the list.

"Boy, that's a hard one," I say. I don't want to mention some dorky movies that she might not like. "I'll start with one then you add one."

"Okay, shoot. Oops!" Linda looks around for any flight attendant that might have heard her.

"Well I don't know about the order but I would put *To Kill a Mocking Bird* up near the top."

"That's one of my favorites too," says Linda. "You remember that scene near the beginning when the little girl who narrates the story is talking to that new kid visiting from out of town? He told her his mom entered his picture in some 'beautiful child' contest and won five dollars. His mom gave it to him and he said he went to the 'picture show twenty times.' I can't believe it ever cost twenty-five cents to go to the movies."

"I remember seeing a dime on the floor in a store I was in a few weeks ago. I looked at it for a moment and left it there. What am I going to do with a dime? Hardly anyone carries cash much anymore, and a dime? It's doesn't even qualify for 'chump change' anymore."

"What's one of your favorites?" I ask, keeping the theme going.

"I liked the movie Titanic. It was romantic and sad at the same time. I don't know if I would put it up there with one of my favorites of all time. It just came to me."

"The Shawshank Redemption is one of my favorites. I liked the way it ended. They both retired on the beach with loads of cash."

Linda can't shake off our earlier conversation. She seems to need "closure."

"What are you going to do with the disc Jake? Are you going to upload it to the net when we get back?"

"Not right away."

"Why not?"

"I want to see what develops first. I know we need to do this but I want to see how things play out for a while; see what they do, if anything. Not long. Just for a few days, a week or so maybe."

Linda doesn't respond. I guess she is okay with my decision.

We arrive in Frankfurt to wait it out for the final leg of our journey back home. We scarf down some airport food and do our best to relax in a not so comfortable setting. Our plane eventually arrives and we watch it taxi over for boarding.

I make sure Linda has the window seat and we sit down. "We only have about eleven hours to go," I say.

"At least we're out of the airport," says Linda. "I was getting a backache from sitting in those chairs."

The plane taxies and lifts off. The flight attendants come round and ask us if we want anything to drink. Linda and I settle on coke and some chips. That should do until the meal is served.

Neither of us is tired yet so we engage in small talk for a while. We speak about what it will feel like to get back to academia, the routine. We agree it will be awkward for a while, everybody pestering us about what had happened, what we had learned. We also agree it wouldn't be appropriate to speak about it in any detail right away. The information will be public soon enough.

I just realize I don't have anything to read but it doesn't matter. I can "read" Linda. I think for a moment about my "list." As it turned out I didn't need the list. There are still blind spots in what we had learned about the tablets and what we thought we knew about how they got here. We might never know.

As Linda and I have come to know each other better we often discuss our feelings on any number of subjects. These exchanges are usually cordial, seldom crossing the line into more sensitive areas, although we do enjoy probing each other's intellectual proclivities from time to time. The moment seems ripe for deeper reflection.

"You told me in Cairo at the club that you kinda-sorta believed in an intelligence operating in the universe. Is that the extent of your belief Jake? Or is there more to it?"

"You're talking more of philosophy now," I say.

"Yes. You said you could take bits and pieces of different philosophies and use what you found helpful and ignore the rest. Isn't that pretty much it?"

"Well when you put it like that, it sounds like I'm window shopping or ordering dinner à la carte. It's a little more nuanced than that."

"So, give me the 'nuanced' version," says Linda.

"Okay," I say. "First off, I don't believe anyone has a lock on how we should live our lives. Some philosophers offer helpful insights, encouraging tidbits about how to resolve conflicts or live better lives. Sometimes they force us to confront truths about ourselves that we would rather ignore, upsetting our expectations as to what we wish to be true. And that can be an eye-opener, allowing us to dig deeper to find a better understanding of ourselves and others."

I look at Linda to see if she is interested or just making conversation. "Go on," she says. "I'm listening."

"Some self-help books tell us what works for them," I continue. "I seriously doubt it works all or even most of the time. They go to great lengths expounding on what has worked in their lives assuming we are just like them, with the same motivations and capacity to achieve similar results. In reality, they are only talking to themselves or others who are already of like-mind but just need a nudge or two to achieve better outcomes in their own lives. To the extent that others of similar persuasion can benefit from such advice, there's nothing wrong with that. The problem is that such individuals are already primed to take advantage of such approaches. It's not a 'one size fits all;' some people benefit, most do not. I spoke earlier of great philosophers and leaders who have given us examples of what is possible to achieve, but basing our happiness or fulfillment upon our ability to emulate their or anyone else's way of life will only lead to disappointment."

"I've enjoyed reading some of the books you're talking about," says Linda. "The trouble with me is, after I've read everything and think I have benefited from the information, I ignore it and go on with my life as if I had never read it. I guess that's because it didn't resonate with me. As you said, it might help people who already feel that way."

Well, at least we agreed on that point. "I don't have a 'template," a written code or set of guidelines I live by. I do believe it is important to continue seeking, searching, examining how I should live my life. Otherwise, it is just as Socrates said, 'The unexamined life is not worth living.' I have examined 'the good life,' climbed the philosophical mountain as it were. I have found that all philosophies or religions have one central tenet in common: They are each seeking to discover the same thing—a path with a heart. Ultimately, as anticlimactic or clichéd as it may seem, that is all any of us can hope to achieve. And those who believe they have arrived, who believe they have found the ultimate answers, are of all people most to be pitied. If you are no longer seeking, no longer striving to discover a better way to live, you have come to the end of your journey. And, contrary to popular opinion, I do not believe it will lead to contentment, for our minds are meant to search for deeper meaning, purpose, knowledge. I do not want to ever complete this journey; it is my reason for living. It is the pursuit of knowledge, of fulfillment, that ultimately guides my life."

Linda has been following my lengthy excursion intently and offers a summary: "So, if you're done . . . you're done."

"Pretty much."

"But isn't that frustrating?" says Linda. "We want answers—well at least I know I do—not a life of pursuing answers."

"You do realize," I counter, "you have just confirmed my philosophy."

Linda looks confused for a moment. "Oh, I got it," she says. "By my picking apart what you just said, I'm challenging what you believe. I'm still questioning everything; still probing, searching—just like you said we should."

I look at Linda in all seriousness: "Of course I could be wrong."

Linda laughs, her warm smile and soft features sinking deeper into my soul. Should I tell her how I feel? What if she doesn't feel the same way? It's a long flight . . . never mind.

"You mentioned you had read the Bible. What was that all about?" asks Linda.

"Cover to cover. The 'begats' were pretty rough."

"That doesn't sound like you," says Linda.

"Nothing sounds like me."

"Is that a koan?"

Summoning up a magisterial tone, I say: "I *am* a koan!"

Linda tilts her head to the side, looking up at me. "Thank you for solving all of my problems Jake," she says facetiously but playfully. "Now can I take a nap?"

"Yes darling, you may go to sleep now." It just slipped out. It felt right, comfortable.

"Goodnight dear," Linda says, as she lays her head back to try and get some rest.

I guess we were both playing, kidding with each other. That's okay. It'll do until we both feel the same way. We shut our eyes and doze off together.

When I wake up Linda has her left cheek on my shoulder, still sleeping. I stay still enjoying our "intimacy." She wakes up, giving no hint that she is embarrassed about using my shoulder for a pillow.

By the time our plane begins its descent, it is approaching eight p.m. After we retrieve our luggage, we hail a cab, not having bothered to make other arrangements before leaving. I turn my phone back on and it is "lit up" with messages. One message is from Syed which says: "Urgent! Please call immediately."

Linda can see from my expression that something is up. "What is it?" she asks.

"It's a message from Syed. He says it is urgent I call him." I figure it would be about five in the morning Cairo time. Syed is probably up. It doesn't matter. I call Syed and he answers immediately. I put the call on speaker so Linda and I can listen together. Syed offers no pleasantries.

"Jake, you must listen carefully. The documents were being taken to another facility for further examination and the car transporting the files was in a terrible accident. The car exploded and all of the documents have been lost. I received a call from the president's office sometime after I left you off at the airport. I could not reach you."

My mind is swimming in confusion. I try desperately to make everything gel, to make sense. It then hits me: It had to happen. We couldn't prevent it. "Syed, are you sure everything has been lost?" Sabotage comes to mind. Could it be a trick? "Maybe they're just trying to fool us—"

"No Jake, I do not think so. I saw the film on television. It was authentic. Of that I am sure."

I don't know what to say. I am shocked, stunned. It couldn't be that simple. It's too pat, too contrived. The "predestination paradox" . . . We can't change the future . . . we can't change the future. I tell Syed I will get back in touch with him shortly.

Linda starts to hash through the implications but I stop her short. "We need to sleep on this. Now is not a good time to think this through. It will be late soon, we're tired. I'll call you tomorrow and we'll see if we can work through it."

"Alright, Jake."

Chapter Nineteen: Predestination Redux

"Things are not always what they seem," so says Phaedrus. Deception can be subtle, deliberately hidden. At other times we simply fail to perceive the obvious, not because we fail to pay attention but because we pay attention to the wrong thing. The tablets contained information that we were not entitled to, not yet. The information was taken from us and rightly so. It wasn't ours; it didn't belong to us. We had no right to it. As I saw it, we made three mistakes: Our first mistake was in failing to recognize that we couldn't alter the future. Our second mistake was in failing to recognize our mistake. And our third mistake was our failure to correct our course before it led to tragedy and the loss of innocent lives. We made all of the "right" decisions which led us inexorably to the wrong course of action. And what of the asteroid? Will we continue to act irresponsibly and thereby cause our demise? Will our attempts to save humanity lead to its destruction?

  I wake up early and sit straight up in bed. The disc! I run through the living room into the den where I had tossed my bags from the night before. I recover the disc and take it to my laptop computer. I place the thumb drive into the computer and wait for the prompt that will allow me to bring up the documents. Nothing. Empty. No documents. Surprisingly, I'm not surprised. Something happened to the disc. Of course. Was it the scanning machine at airport security? Did I do something to wipe the disc clean? Did little green men . . . It didn't matter; it was inevitable.

I call Syed but don't tell him about the extra disc I had made. What for? It wouldn't make any difference. Syed says he received a call requesting him to come into the local office of the Ministry of International Cooperation.

They have a few questions for him.

"What is that all about?" I ask, my paranoia kicking in again. What could they want with Dr. Syed? That's not their CIA, is it?

"It must be about the documents," says Syed. "I do not know what their intentions are but I am sure it would be best for me to cooperate."

I can't detect any apprehension in Syed's voice. Of course, that tells me nothing. He might just be hiding any concerns he has. He says his appointment is in a couple of hours. I think it best to wait until I know what it is all about before I draw any more conclusions. "I agree Syed. It's best to cooperate and see what they want." I ask Syed to call me after the interview and let me know how it went.

Linda is expected back at university tomorrow. I call her and suggest we meet for coffee at the Med Café on Campus Drive at Stanford.

I arrive at the café and order two coffees while waiting for Linda. She arrives looking cheerful, upbeat. "Looking forward to getting back to the grind?"

"I am," says Linda. "I need to get oriented again; get back on track and put everything behind me."

Linda's comment is surprising. "I didn't think you would be so eager to get back to your duties."

"I enjoy teaching," she says. "It keeps my mind sharp and I get to continue my research. I don't know what to make of everything that has happened. I have a ton of messages asking me what happened, what I know, and on and on. I've answered some of them but I don't know how to respond."

"Same here," I say. "I haven't gotten back to everyone yet either. I wonder what they want with Syed."

"What do you mean?" asks Linda. "What's going on?"

"Some people want to speak with Syed about the documents. I don't know where they're from. He said it was some agency of international cooperation or something. I don't know. It could be they're just using the location to interview him and they are connected with some other department."

"You don't know who 'they' are?"

"Somebody in the government; that's all I know. I asked Syed to call me after his 'interview' and let me know what's going on. It could just be routine."

"What about us? Do you think they will want to talk to us too?"

"Probably; it makes sense. The documents contained a great deal of scientific knowledge that they would want to get their hands on. I can't blame them."

"Let me know what he says when he calls," she says, taking a sip of coffee. "Ah, good old-fashioned coffee. I guess Syed feels the same way about his Turkish coffee."

"I think he is more of a tea drinker. But you're right, we are creatures of habit." I can't put it off any longer, I have to tell her. "The copy of the disc I brought back from Cairo is blank. I checked it several times and there's nothing there."

Linda looks like she was expecting the news. "That figures," she says.

Whether Linda is entering some new Zen phase or just fed up with everything, I can't tell. My assessment of what our journey to Cairo would mean for the future is bleak. I am hesitant to share my thoughts with Linda as I can't think of anything positive that might come of it. Getting Linda's view would invite a discussion about the darker implications of what our involvement might mean for the future.

We talk for a while longer, mostly about Linda's plans for the rest of the semester. We avoid any further mention of Syed's call and his pending "interrogation." We say goodbye and I tell Linda I will call her as soon as I hear back from Syed.

I go home and sit out on the back patio. It feels good to be home and have some time to relax and regroup. Syed calls, sounding like his usual self, calm and composed.

"Jake, my friend, I have some good news. It is a little disappointing but I believe it may explain a great deal about what has been happening."

"What did they want? What did they say?" Syed is making it sound so simple, routine.

"It would appear we have made some incorrect assumptions," he says.

I'm starting to get impatient. "What are you talking about, Syed?"

"They explained to me everything. The documents are from your country. They are connected with much research that is being conducted at the, I believe he said, the Santa Fe Institute in your state of New Mexico."

"From us? They're from us? That doesn't make any sense—"

"It was explained very carefully to me," he continues. "The institute has been working on advanced algorithms to determine the course of future scientific research. The documents are the product of their continued effort to complete the present investigations that are being conducted in the scientific community. They are 'projections' and are based on current research in many fields of scientific inquiry."

This is bewildering, bizarre. I close my eyes tightly, trying to squeeze out some space to fill up with an intelligent response. "But that just leaves more questions unanswered." I can think of several right off the bat.

Just then Syed's phone intercepts an incoming call. It is from the agency that had brought Syed in for questioning. I tell him to go ahead and take it and call me back. I don't feel like waiting on hold.

As I am waiting for Syed to call me back, I start to think about what he had told me so far. I need to organize my thoughts before Syed contacts me again. I go through what I call my "laundry list"—a mental approach that tickles through every detail I can think of until I am satisfied I haven't missed anything.

Several questions come easily to mind: If these documents, the information, originated in the United States why would they be sent to Egypt of all places? Why were they encased in "Egyptian tablets"? And what of the technology that allowed the information contained on the miniaturized discs to be brought up simply through a scanning process, even if the scanning machine was state of the art? And why couldn't we detect how the inscriptions were made? How could they have fooled us about their age and authenticity? The information isn't just sophisticated—a mere "projection" of possibilities—it is well thought out and finessed. It seems more a finished product than "speculation." And the discs—what happened to the information? The discs we took to the university for our rehearsal and the disc I took back with me—vanishing in both instances? Sorry, it doesn't jive; something is out of whack. Was Syed being told a "cover story"? What am I missing?

Syed calls back. He tells me the agency had been "mistaken" about the Santa Fe Institute's involvement. It was an "international colloquium." The details are unimportant, or so he says.

I'm not buying it; not a chance. I could argue with Syed, ask him how he could have been duped so easily. I don't. I thank Syed for the update.

I can't imagine what Linda will make of this new bombshell. I send her a text asking her to call me and she does so right away.

"What did Syed tell you?" she asks.

"He said the documents were part of an international effort to create 'projections' of current research to show where the ideas might lead in the future. He first said—"

"You don't believe any of that do you?"

"No, of course not," I say, glad to know Linda is on board with my thinking.

"But Syed bought it?" asks Linda, incredulous. "I don't believe it. They're hiding something. Syed wouldn't have fallen for it. I wonder if he was worried about someone listening in on the call. And what about us? Do you think they are listening in now?"

"I wouldn't doubt it but I don't know what we can do about anything now. We no longer have the information, but I am convinced they do. They're lying to us but we don't have any way to prove it. We don't have anything to go on. The agents sent to transfer the documents were sacrificed to throw us off track. They want us to believe the documents were destroyed so we'll just give up. We should have released the information when we had a chance."

"It doesn't even hold together," Linda says. "Their cover story is totally bogus." Linda pauses for a moment, then lets out an audible sigh. "I guess you're right. I don't see how we can prove anything."

Linda is all gloom and doom. I don't blame her. I feel the same way. It does look hopeless. And what is going on with Syed? I wish I knew if he believed any of what he told me. He sounded like he did.

"How about dinner tonight?" I say.

"I want to be fresh for tomorrow when I get back. We could meet at Starbucks this afternoon. How about four o'clock?"

"Sounds good. Maybe we can make some sense of this 'doublespeak' after we have stepped back for a moment to give it some reflection. See you soon."

I feel like calling Syed to see if he will retract his former position. I just cannot believe Syed, a man of calm reflection and sound mind, could be steamrolled by a government bureaucrat concocting such an obviously false narrative.

I always reserve a modicum of doubt when analyzing my personal opinions about scientific matters. The phrase, I could be wrong, has become my mantra for as long as I can remember. When I first read about Einstein's mistake in adding a "cosmological constant" to counteract gravity and thereby hold the universe static, it hit me like the proverbial "ton of bricks." If Einstein could make such a mistake, who am I to be allowed an exemption? Never mind the fact that we ended up using the lambda term for the energy density of the vacuum of space. Einstein's original application was inserted in his formulation for the wrong reason. It was clearly a mistake.

Am I making a similar mistake? As preposterous as the story was the agency told Syed, is what I want to believe any more plausible? Us, in the future, "sending" information back in time, ensconced in tablets from ancient Egypt, discovered serendipitously by a nefarious character and brought to the attention of none other than the esteemed Dr. Syed. As a scientist, I know full well that, whatever scenario turns out to be true, I have to remain open-minded. Marcellus' comment to Horatio springs to mind: "Something is rotten in the state of Denmark"—make that Egypt.

I try to relax and put the matter on hold until Linda and I get together later this afternoon. While checking out the Santa Fe Institute on line I receive a call.

"Hello, is this Dr. Jake Banner?"

"Yes, who is this?"

"Oh hi, I'm Dr. Theresa Adams. I am the coordinator for The Directorate of Science and Technology. I wonder if I might be able to speak to you for a moment."

"What is this about?" I ask, apprehensively.

"We just want to have a word with you about your recent visit to Cairo. We have a few questions, mostly routine. Would that be alright?"

I've never liked the word "we" when speaking to one person. "What is it you want to know? Couldn't it be handled over the phone?"

"No, I'm very sorry but there are certain procedures we have to follow. It doesn't have to be anything formal. I spoke to Dr. Cooper from the university and she agreed to meet with me at one o'clock. Would that be alright with you?"

An incoming text from Linda arrives. I'll look at it in a minute. I wonder if it's about this lady I'm talking to right now. "Where would you like to meet?"

"Well I'm just in town for a little while but we could use the offices at the U. S. General Accounting Office on Howard Street if that would be okay. Dr. Cooper agreed to meet with me there. Do you know where it is?"

"Yes, I can find it. I'll see you then." I hang up and look at Linda's text message. She wants me to call her about our meeting. I call her right away.

I know Linda will want to get straight to the point so I beat her to the punch: "I think it's a whitewash and they're going to try the same thing on us they did with Syed."

"Well I suppose we could try and be heroes but what's the point," says Linda. "They have all the ammunition, the information. We got nothin."

"I think we should just play along and pretend we believe them. No sense in testing the limits of what they might be capable of. Do you know anything about the office she's from? I don't think I've heard of it. This is the first time I've ever received a call from them."

"I look them up on the net. They're connected with the CIA in some capacity having to do with classified research and science technology." Linda lets out an audible sigh. "Jake, I don't think they're going to tell us anything. They just want to know what we know while holding all the cards."

I can't argue with Linda's logic. "Yep," I say, "that's pretty much the way I see it. The cookie has crumbled."

"After we meet with her let's recap where we stand. I'd like to get things behind me the best I can. I hate going back to university tomorrow with all of this hanging over my head. At least we'll know where they're coming from."

We both agree. I tell her there is no use in taking two cars to the meeting so I will pick her up around twelve thirty or so.

I go in the house and turn on the television. It doesn't matter to me which station. I seldom watch TV. I just need some background noise. It sometimes steadies my thoughts; something to do with my wiring.

Linda and I speak with Dr. Adams in a conference room at the General Accounting Office. The "interview" goes as we suspected. The administrator asks if it would be alright to tape our conversation. "I'm terrible at taking notes," she says. After peppering us with questions about how we came into possession of the plates and what eventually happened to them, she thanks us for taking the time to see her. She seems satisfied with our answers and doesn't lean on us for anything more than we are willing to tell her. I ask her if the documents if they were the work product of an international effort. She says she isn't authorized, blah blah blah.

After our meeting, we go to a nearby café and order coffee. Linda has a look of resignation. "I wish I could talk to Syed but I don't want to challenge his interpretation. He's entitled to his opinion. I didn't get to know him well but he doesn't strike me as the sort that would believe such nonsense. Who knows, maybe he just didn't want to say anything against the government's argument, lame as it is. You know him better. Do you think he believes what we're being told?"

"No, I don't. And you're right, I do know him better. Well enough to know he wouldn't be taken in by the bureaucratic tripe we're being fed."

"So why is he playing along?" Linda asks, immediately catching the contradiction. She then answers her own question. "Well, he's just doing the same thing we are. What's the use of arguing with the government anyway? He probably just spewed that nonsense in case anyone had been listening in when he called you."

My paranoia and conspiracy have finally infected Linda. Welcome aboard! I knew they were out to get us! I mentally bring up my "laundry list" and go through each item with Linda to see if she can come up with any explanations.

"There is no credible explanation for why the tablets would be in Egypt," Linda begins, "or be 'camouflaged' the way they were. The way the discs brought up information embedded in limestone was uncanny. I don't think we're quite there yet technology-wise."

"What did you think of the documents you reviewed?" I ask. "Did they seem like the real thing or just 'projections' of possibilities?"

"If we have a way, by some monstrous algorithm or otherwise, to come up with the sophisticated information I saw, why would we be sitting on it? I don't believe it is simply guesswork. The documents had a level of comprehension of subject matter I would never have expected from any think tank. I believe it is authentic. What were you're impressions?"

"I feel the same way; they're genuine. Losing the information on the discs is what baffles me. The discs we brought to the Cairo University and the extra copy I made were all wiped clean. I'm not saying it couldn't be explained. It's technology. I guess it could happen, data does get corrupted, but it is one heck of a coincidence."

"Even if we concede your point about the information on the discs," says Linda, "that still leaves a lot unexplained." Linda finishes her coffee. "I better get home, if that's alright."

"Sure," I say. I drop Linda off back at her place and then head home. Time to recharge. There's nothing like a power nap to recalibrate the neurons.

I wake up and check the time. It's only been ten minutes. Asteroid! What about the asteroid? Sometimes I have a thought and I know it is important without knowing why. Why was I thinking about the asteroid just before I woke up? What about it? It's not due for several years from now. Why is it so important? And then it hits me: If we could find out when the information about the asteroid's impending course change could have been known that would provide a clue as to when the chart was generated. If there is no hint whatsoever that any course change was even possible to track based on current information, that would put the document's creation much further out in the future. Otherwise, if it could have been detected based on what celestial mechanics can *presently* tell us that would put the document closer to our time. This makes sense. Now how do I check it out?

I send Jeremy a text asking him to give me a call. While I wait for him to get back to me, I call the Santa Fe Institute in New Mexico. It's a longshot but they might tell me something. They didn't. My credentials meant nothing to them. They thanked me for my interest and gave me the brush off. I could pursue the matter further but what's the point.

Jeremy calls me back.

"Hey, you made it back," he says, sparing me the embarrassment of having to explain why our world-shattering press conference was canceled.

"I need your help, Jeremy. There's something I'm trying to figure out but I'm not sure how to go about it."

"Yes Jake, black holes are portals to other universes. I worked it out last night so I would be ready with the proof when you got back." That's Jeremy, the penultimate intellectual buffoon. Which explains why he and I like The Three Stooges. I consider an appreciation of the Three Stooges the litmus test of a highly intelligent or profoundly stupid individual. We like to think we just make the cut on the high side.

"Seriously Jeremy—"

"Sorry Jake, go ahead. What have you got?"

"Let's say an asteroid was on track to pass by Earth in the future but we discovered another object or some anomaly that might interfere with its path. Would we, I mean you and me or anyone else who might be interested, have access to such information? Or would it be kept sequestered where only a few insiders would know? What I mean is, would the public be aware of it or would the 'guardians of the universe' keep it secret?"

"If it were imminent everyone would know about it. The amateur astronomer class wouldn't let the astronomical or astrophysics community get away with keeping such information quiet. Of course, if we discovered evidence about some celestial event quite a ways down the road, we might keep it on the QT until we knew more, but even then, it would leak out sooner rather than later. You can't keep information like that quiet for very long."

"If I gave you a specific asteroid would you be able to find out if there is any projected adjustment in its flight path or not?"

"I may not be able to get the information right away—"

"That's okay, but you think you could get it, right?"

"There are only three degrees of separation from me and NASA," says Jeremy. "I'm sure I could find out. If it has to do with an event that is far off into the future, I don't think I'd be stepping on anyone's toes if I looked into it."

I have no hesitation in giving Jeremy the specific information about the asteroid. I don't see any downside and it just might help me to know more about the timeline of the documents. He says he will use his "network" and let me know what he finds.

I sit on the back patio for a while reading, trying to relax and put things behind me. I know I have to get on with my life eventually. Unless Jeremy comes up with something concrete, I don't see much hope in pursuing the matter. I eventually go back inside and sit down in the living room.

My thoughts turn to my trip to Cairo, how Syed and I had become close friends, how my feelings for Linda have deepened. Linda will be back at university tomorrow, soon reabsorbed into her academic routine. Her memory of our adventure together eventually fading; not right away but soon enough. The specific would become the general, the general, in turn, fading into a recycled version of events, continually reconstructed and redrawn in her mind's eye until all that is left is a vague notion of time, place, and history.

When I want to put things into perspective I sometimes think about the big picture. We live in what scientists refer to as a "light cone." We are in the middle, the observer, and in opposite directions stretch two light cones—one projecting into the future and the other into the past. I see the cones as representing the maximum time-frame within which any other persons may know of our existence: Our parents, grandparents, and others born before us who have passed on during our lifetime and those who knew us while we were alive and survive us. When these two sets of individuals are gone, the living memory of us also vanishes.

Is it any different for the human family as a whole? Eventually, there will be the last generation, perhaps far distant in time and place, beyond the confines of Earth even. When that last generation vanishes the remembrance of them will be forgotten. In the final analysis, will the epitaph of the human family—or even the universe—really matter? If there is no light cone left there can be no remembrance of . . . anything.

I call a cab and go to a local pub in Palo Alto. I have one objective: to relax, escape, and think of . . . nothing.

Chapter Twenty: Stranger than Truth

Linda calls me from the university. Syed had called her and asked if she was on her office phone. She told him she was at another work station when the intercom came through. He then told her why he had played along with the government's version of events and that he suspects a cover-up.

I knew Syed wouldn't fall for it! Linda said he wanted to talk to me and requested I take his call at the university as Linda had. Linda asks me if I can come to the university around noon so Syed can contact us in the evening Cairo time. Maybe the intrigue isn't over after all. Clandestine operations, secure communication lines, government subterfuge. I can't wait to hear what else Syed might have learned. He is closer to the action. Maybe he has gotten a scoop, cracked the case wide open. It is a grand scheme, after all, a government plot to silence the truth. The CIA wants to close the door before we discover what they are really up to. I should have placed more faith in my paranoiac and conspiratorial mindset.

I reach the university just before noon and speak with Linda for a few minutes before she has to leave. She isn't able to provide any more information that we both already know but gives me the impression Syed might have more to say.

Syed calls me and reiterates his position regarding the government's story about the documents. He then provides another piece to the puzzle.

"Jake, about seven or eight months ago an Egyptologist working at the site in the Valley of Nobles in Luxor called me about a find. It was a tablet, normal in all respect as far as he could tell. But there were two features he thought might be of some interest: the date of the tablet was very old and the writing, although very similar to the Egyptian hieroglyphs that we normally analyze, was different in some respects. I thought nothing of it at the time. It was not so very unusual. We do find artifacts that differ in some ways from what we are accustomed to finding."

"Do you think it might mean something?" I ask, eager to discover where this new bit of information may lead. "Could it be connected to our discovery?" I'm sure Syed would not mention the find if it held no interest beyond mere curiosity.

"It may provide a clue, Jake. I do not know. It is a possibility."

"How old is the tablet?"

"We are not yet certain. They have made a preliminary estimate but an assay has not yet been performed. Deciphering the writing is of most interest for now."

If this find is connected somehow it should have a similar date to the ones we discovered. "How old do you think it is? What is the preliminary estimate?"

"The writing is somewhat similar to the period we have estimated for the tablets we discovered, but we cannot be certain until we have examined it further. I do not know if it is of any importance to us Jake but I thought you would want to know."

I think of the cover letter referring to an "alternate timeline." Does this plate have something to do with our discovery? Could it provide a missing link, the clue we need to put the pieces of our puzzle together? "Syed, I think it is important to find out how old the tablet is. It might shed some light on our discovery. Maybe not, but it is worth checking out, don't you think?"

"I do not have authority on the dig. It has been some time since I was involved in the project and I have not played a major role in the work being done there." Syed pauses for a moment. "But I will see what I can find out. It will give me an excuse to visit Luxor and do some snooping. I will let you know as soon as I have anything to tell you. Perhaps I can provide a reason for another visit."

"It's your turn to visit me," I say. "Of course, I don't live in a palace but I'm sure I can make the accommodations comfortable."

Syed laughs. It is good to hear from him again. I send Linda a text telling her I will call her later and fill her in on my conversation with Syed.

I have always hated "loose ends." If I'm working on a project or doing research and I'm missing some information or lacking critical data it drives me crazy. I don't have the luxury of an Einstein to simply "plug" an equation to make it balance, thus tailoring my formula to generate the desired outcome. I have to grind it out, stay the course until I find a solution. This is no different. If I put my mind to it, put the pieces together, I can make sense of everything. I can do this. I need to work with what I know leaving nothing out, however trivial it might seem. There is always a key to every problem, a salient fact, or surmise that ties everything together. I just have to find it.

Any acceptable explanation will have to admit of a solution outside mainstream scientific orthodoxy. I make a mental summary of where things stand so far: We don't just stumble upon or "discover" tablets from ancient Egypt containing advanced technology. The documents are more than mere "projections" of present knowledge. Until Jeremy checks out the data on the asteroid, I can't draw any substantive conclusion about what role, if any, it might play. Syed should be able to provide more information about the find at Luxor. As far as the governments' attempt at a cover-up, it only confirmed their duplicity. The only thing left to consider is the cover letter referencing the documents. What is it trying to tell us that we are overlooking?

At first, I can't remember anything except the binary notation showing what we thought were the first three capital letters of the alphabet. There was something else . . . the timeline. What about it? Alternate. It refers to an "alternate timeline." It didn't make sense when I read it. Do I have enough to go on now? An alternate timeline usually means we are looking at two or more paths through time. It could be something as simple as two people encountering different life-trajectories or "world lines," different histories in ordinary space and time—nothing remarkable. Or, it could be suggesting something different, quite different: parallel universes where different versions of us play out different life scenarios in alternate universes similar but not identical to our own.

My analysis eventually brings me to an idea of "many interacting worlds" put forward by the *Griffith's Center for Quantum Dynamics* and the University of California. In this interpretation parallel worlds interact using a repulsive force. The paper presenting these ideas was published in the *Physical Review X* which carries substantial weight in the scientific community.

So far, determining whether or not such possibilities exist requires "our" universe to collide with another one, leaving behind an alteration in the cosmic microwave background. If some parallel universes are similar or even identical to ours and capable of interaction—a very big "if"—it would offer an intriguing possibility for explaining how we came into possession of the tablets.

The cosmic microwave background! My heart races! I feel like Einstein must have when he had "the happiest thought of his life." He had discovered that, from the standpoint of a freely falling observer, gravity is non-existent. And, conversely, an observer in an elevator could not confirm whether or not her weight was due to acceleration or gravity.

The shift or wiggle in the image on the plate had been screaming at me and I was too distracted to notice. There is no accounting for the degree to which supposedly intelligent scientists fail to grasp the obvious. This "jiggle" could represent interference in the background radiation due to any number of causes. If the multiverse theory holds the shifted image could represent an exchange of energy between universes, due either to a collision or interference of some kind. Such an incident would have happened long ago and therein lays the rub. What possible connection could there be with a collision happening long ago and relatively recent events? Could a collision between universes—even if removed from the present by unfathomable time—leave a "cosmic seed" that captured more than the cosmic background radiation; perhaps a "blueprint" encapsulating future events?

I feel like a vaudeville performer juggling plates on sticks: "Many worlds," "parallel universes," "multiverses," "colliding universes." It's easy to slip and speak of different theories as if they were the same, maybe they are at some deeper level. I am no longer certain if I am keeping things straight in my mind and I began to develop what my optometrist calls a "soft migraine"—the gentle glow of white lights unaccompanied by a headache. I usually get these migraines when I consume too much sugar or concentrate for too long.

The image we now have on the plate would simply be an updated reading of the cosmic background radiation occasioned by newer technology or improved methods of resolution—faint perhaps, but visible nonetheless. The "double image" on the plate may have been exaggerated to highlight the newly discovered anomaly.

Disregarding the multiverse theory's implication for free will, I focus on what matters now: the tablet showing the cosmic microwave background, the radiation left over from the Big Bang. The distortion or whatever it is of the image on the tablet is the "smoking gun."

I have photos of the tablet. I'm sure Linda and Syed do as well. Why hadn't anyone from the government agencies asked us if we had any photos? Who knows? More bureaucratic incompetence. But will our pictures be enough? They're just pixels. Will the resolution be clear enough to make a genuine comparison with data currently on hand? Perhaps we could enhance them somehow if needed. I do a quick comparison on my own, pulling up the latest image I can find of the background radiation. The difference between the two images is plain. But how can I know for sure what it means?

There is more than one version of a multiverse. The type of multiverse I'm considering would result from inflation, a theory explaining the universe in its first few moments. In this scenario "bubble universes" arise when individual regions of space that were originally expanding stop inflating.

I don't believe the double image I'm seeing on the photograph represents a temperature fluctuation or signature of a bubble collision. The dynamics of these collisions require computer models and are extremely complex but they don't usually result in an image like the one on the tablet we had uncovered. String theory is of no immediate help as it doesn't offer a unique solution that might explain the anomaly. The more I think about it the more the possibility strikes me: The universe may not make sense, at least not in a way that we can understand without altering our view of reality.

When I call Linda latter in the day, she is intrigued to hear about the tablet Syed had mentioned to me, the discovery in Luxor. He had not told her about the tablet but she didn't seem to be upset that he had not confided in her. She probably just chalked it up to the closer bond that Syed and I have.

"When do you think Syed will know more about the tablet?" asks Linda. "Do you think it could be connected in some way?"

"I don't know how long it'll take. He mentioned he would be visiting Luxor where the tablet was first located. It's the age of the tablet that I'm interested in. If it is the same age as the ones we had it might tell us something. He said the hieroglyphic writing is similar but not identical to other writings from the same period. I wonder—"

"Do you suppose that's where the other tablets came from?" Linda sounds distracted. Transitioning from Cairo and a world of intrigue to academia couldn't be easy.

"Hmm, I hadn't thought of that. That would tell us something. I don't what exactly but it could be significant. I'll call you as soon as I know more."

Jeremy calls and tells me there isn't any current information available suggesting any foreseeable change in the asteroid's telemetry, no anticipated interception or anomalies have been detected. I thank Jeremy for his help and end the call.

That leaves the cosmic background signature and the tablet Syed was checking on to provide a clue, any clue. As for the distortion in the image of the background radiation, it seems clear that it represents an adjustment or juxtaposition of the background image. Could an experienced cosmologist provide a specific cause or reason for the distortion in the image, explain how it came about? Working out the solution to this conundrum is up to me, Linda and Syed. The tablet uncovered at Luxor is the only possible remaining link. If it is connected in any way to the previous find it should become apparent if it can be properly interpreted and dated. Why hadn't Syed mentioned the tablet earlier? Perhaps the time and place of its discovery put it out of Syed's thoughts at the time of my visit. It might not be related to anything we're investigating. I'll know soon enough.

I have other matters to attend to as well, though perhaps they can wait. I want to complete a research project I had been working on before sabbatical's end, "earning" my time off. And then there is the matter of my postponed vacation. Vacationing in the Caribbean seems like a petty thing to be thinking about now. As for my research, if the documents I had reviewed are genuine it would make my project obsolete.

Over the next few days, Linda and I keep in touch by phone and text, have lunch together a couple of times. I haven't been able to muster the courage to take our relationship further. I am still harboring the notion that Linda doesn't see us as anything more than friends. I will never know for sure until one of us makes the next move. I guess it will have to be me.

Just after noon on the following Tuesday Syed calls to give me an update on his trip to Luxor.

"Jake, it is good to talk to you again."

"Syed, how are things in Luxor?"

"Hot, very hot."

"So, what did you find out?"

"It is very interesting Jake. The hieroglyphs have been deciphered and the tablet has been identified. There is a problem, however. The tablet is identical in every way to one that has already been uncovered in the area. It is only the hieroglyphs that have been very slightly altered from the original. The writing on the tablet has an odd mixture of Mesopotamian cuneiform. Some Egyptologists believe hieroglyphic writing developed from this style but it remains disputed. The best way I can explain it is to compare the writing to what we would call an accent when someone speaks in a second language. The variation between the two tablets is noticeable but the differences are minor. It did not take so very long to translate between the two."

"That is strange. Were you able to determine its age?"

"Radiocarbon dating places the tablet at approximately ten thousand years in the past. That would place the tablet near eight thousand BC. The interesting thing to note, Jake, is the nature of the find. It is known as the Ivory Tablet of Zet associated with the first dynasty and fourth Egyptian pharaoh. It was recovered in High Egypt in the north-west sector of Luxor. The tablet is an older representation of a tablet we had uncovered over a century ago."

"When was the pharaoh living?" I ask, trying to pinpoint the timeframe.

"His reign was during the latter half of the thirty-first century BC."

"Was there anything special about his reign?" I ask, still searching for a clue to help me piece things together.

"We believe the queen was his sister Meredith and that she may have also ruled as a Pharaoh when he died. He may have had another wife and so on but nothing that would be of tremendous interest or importance to us. His fame is mostly due to the discovery of the many refined steles which have been preserved very nicely."

The bizarre is now becoming commonplace. I attempt to summarize Syed's findings to make sure I have it right. "So, we have two tablets, identical in most respects, separated by approximately five thousand years. The only difference you find is in the writing; but otherwise, it is the same information about King Zet's reign. Is that right?"

"Yes, that is correct." Syed pauses for a moment before letting the other shoe drop. "There is one more wrinkle we should consider."

Who does Syed think he is! Colombo? Just tell me! "What is that?" I ask, calmly.

"The original tablet is missing. We have ways to compare the plates of course. There are many detailed photographs and replicas of the ivory tablet. But it is quite odd, is it not?"

Yeah, it's that. I have to concentrate, to get more information from Syed. There's got to be a connection. "When did it go missing?"

"I do not know the exact date. It was last year I believe, during the summer. I could find out very easily if you think it might be of some help."

"It might. What about the tablet you have now? When was that uncovered?"

"Oh, I can check that right away. Just a moment."

Syed looks up the information and gives me the date: "July thirtieth of last year."

"Can you find out the date when the other plate disappeared?"

"Yes, of course. It is very late now. I will make some calls tomorrow and let you know. Do you think it may be important?"

"It could be," I say. I don't know if it is important or not. I just want every scrap of information I can get.

Syed and I chat for a while longer. He tells me there is an American movie he wants to see, *Lawrence of Arabia*. I am surprised he hasn't seen it already, being the Sharif aficionado and all. I can just imagine watching the movie with Syed: Omar Sharif, Cairo University alumnus, playing the part of an Arab coming to the aid of British Lieutenant T.E. Lawrence played by Peter O'Toole, helping the British fight against the Turks. A good movie by anyone's standards but the politics would drive Syed crazy.

No matter what is going on in our lives we continually seek "normalcy," the feeling that everything is on an even keel. We crave excitement, fear danger, and, when our lives are turned upside down or we encounter disruptions to the status quo, want nothing more than to "go back home" to the familiar. Am I permanently leaving the "familiar," uncovering a world so far removed from my understanding of reality that it is making me uncomfortable? I am discovering a universe far different from the one I grew up in, a new world in which information is unstable, perhaps subject to manipulation. The prospect of living in such a universe is unnerving. I find it difficult to keep grounded, to keep my thoughts from wandering, speculating about what it might all mean.

The following morning Syed sends me a text: "The date the tablet went missing is the end of July. The exact date is most likely the 29th or 30th of the month. Let me know if you make any progress." That was it, the full text. Syed hadn't drawn any conclusions. The ball is in my court.

I don't know what I can find out about the tablet containing the double image of the background radiation. I always have Jeremy to count on. He is probably suspicious about my line of questioning so far. What would he think if I asked him about the image of the cosmic background radiation? I know Jeremy well enough to know he is too smart to be fooled. If I bring up yet another "anomaly" for him to help me investigate, he will surely put two-and-two together. No doubt about that. At this stage in my unresolved search for answers, Jeremy's suspicions would be welcome. I send Jeremy a text asking him when would be a good time to see him. He replies that I can drop by between three and three-thirty this afternoon.

Jeremy has a small office on campus where he keeps a computer, assorted books and stacks of papers strewn about, many on the floor. His office reminds me of an account I read about Alan Guth's office. Alan is a theoretical physicist and cosmologist at the *Massachusetts Institute of Technology* in Cambridge, Massachusetts. A visitor once reported that the floor of Alan's office was layered with unopened mail. The person relating the account picked up an envelope from the floor at random. The postmark was ten years old. I always keep my work and home environment clean and organized; whereas Jeremy—let's just say I would be "Felix" and he would be "Oscar." Although not afflicted with hypochondria I have other disorders that make up for it.

"Hi, Jake," says Jeremy. "What mystery have you brought me today? No wait, don't tell me. You've discovered a wormhole and want verification. Am I right?"

I know I have to get past these barbs before Jeremy will get serious. It is just his nature; you either love this part of his personality or, like me, try not to mind so much. We can both give and take on occasion and, usually, I appreciate his sense of humor. Today, I want answers.

"How'd you guess?" I say, playing along.

"Really?"

"No, not *really*. I need you to look at something and tell me what you think." I show him the picture of the background radiation from my cell phone.

"Huh," he says. "That's weird."

"What?" I suppose he means the "double image" that Linda noticed when she first saw the tablet.

"Let me do something. Hang on a minute. Do you mind if I upload the photo?"

"No, go ahead." What does he have in mind? I haven't a clue. But I do know if something is out of kilter Jeremy can figure out what it *might* be.

"Latitude thirty and longitude thirty-one . . . I guess that would put you smack dab in Cairo when you took this picture."

"You can tell that from the picture?" I say. Then I remember the book I had read on the plane.

"Piece of cake," says Jeremy. "I can even enhance the photo to find out more about—"

"Never mind. What do you make of it?"

"Let me pull up another program," he says. Jeremy looks at the computer screen for a moment as if he can conjure up the file by looking at it. "Oh, I know where it is. Hang on a second."

Jeremy brings up a different program and loads the photo onto the page. "This will allow us to look deeper into the image and see what else we've got. It's not exactly 3-D but it can accomplish pretty much the same thing. Let's see . . . yeah, that's what I thought. This image looks like a hologram. There are pixels embedded in pixels. Cool!"

Hey, that's Linda's word. He has no right to use it! Well, I guess she doesn't have a patent on it. "How can you tell that just by looking at the image? Couldn't it be something else; a distortion in the picture maybe?"

"It could be, but if it is, it's got more information than a flat two-dimensional image should have. Look at this." Jeremy points to an arbitrary spot on the picture showing what I take to be a double image from the exposure.

"What am I looking at? It just looks like a double image to me. What do you see?"

"Look at the difference in resolution," says Jeremy, trying to get me to see what is obvious to his trained eye. "Do you have a . . . never mind." Jeremy reaches into his pocket and pulls out his wallet. He shows me a credit card containing what he calls a *security hologram*. "This isn't a hologram in the strictest sense but it gives you the idea. The images are 'stacked' on top of each other giving you alternating views depending on your angle of sight." He then moves the card back and forth so I can see the images alternate. "The subject is superimposed or sandwiched between two layers, the image in the middle on top of the one in the background." Jeremy looks up at me. "Have I lost you?"

"No, I think I got it." I had seen this kind of thing before. I am beginning to see what Jeremy means. Still, it could be a bigger leap than warranted. All we have is a picture and the picture isn't holographic. If anything, it is a picture of a hologram. How could anyone tell for sure what we are seeing?

Jeremy and I discuss the "holographic principle" and how it might relate to the image we are seeing on the picture of the background radiation. The theory tells us that the universe, which we see as consisting of three dimensions of space and one of time, is a two-dimensional construction coating the cosmological horizon. We observe three dimensions at macroscopic scales but our impression is engineered by the two-dimensional information array embedded on the boundary.

The "holographic principle" resolves the information paradox associated with black holes. Essentially, the information content of objects falling into a black hole is preserved in the oscillations on the surface of the event horizon. Analogously, the information content of the universe might lie on the "event horizon"—a stretch and mischaracterization of the term—or boundary of our universe.

On my drive home I think about how odd it is that we take so much for granted about the universe we live in, our home. Exploring the universe is "somebody else's job." It's as if someone moves into a large, new home that she's never before seen. Immediately after moving in she piles all of her belongings onto the living room floor and makes this *one* room her entire living quarters. Not once does she leave the room to explore the rest of the house.

Most of us, most of the time, never bother to consider what lies beyond our little "neck of the woods," our "kith and kin." But there is more, much more to our cosmological neighborhood, beyond our galaxy, stretching out to infinity in all directions. The universe is my home and I live within a dust mote in a tiny, dark closet shut off from the rest of an even greater mansion; one so immense that it is impossible to imagine. I want to explore it, see what other "rooms" there are. If I could escape the Milky Way, plowing my way through the "Local Group"—a hodgepodge of some thirty galaxies or so—streaming past Andromeda and the other galaxies until I break through the intergalactic sheets, threads of galactic material separated by immensely large voids of "empty" space, where would I end up? Back where I started? Maybe . . . maybe not.

If I could "roll-up" the fabric of the multiverse—squashing the space created between each region by inflationary expansion—and pull it toward me until the last edge of the multiverse came into view, what would I then see? I have no doubt . . . none whatsoever . . . that I would see new worlds so far removed in time and space that the fundamental or "effective" laws of physics would no longer apply. The very word "space" would be obsolete, meaningless.

I am reminded of a paper I had written as an undergrad when I had the opportunity to take a philosophy course. It wasn't a requirement for my major but I was interested in the course and thought it might be interesting. My subject was Parmenides, a Greek philosopher living in the fifth century BC. Very little survives of his work, a single poem entitled *On Nature*. In one part of the poem, he writes that reality is timeless, changeless, uniform. What I took away from my study of this Greek thinker was profound but simple to state: Nothing but being can be. And the corollary—non-being cannot be. We, therefore, conclude that there can be no multiplicity of being as there would be no separator. There is no such thing as non-being? Not possible. I think of the now-dated joke about the Zen master who asks the hotdog vendor to "make me one with everything."

I call Linda and reach her just after she arrives home.

"Hey," I say, customarily.

"Hey yourself."

"How about dinner this evening?"

"It's a school night," says Linda, sounding lighthearted at the end of the day.

"Oh c'mon," I say. "My treat. I'll pick you up at seven."

"Okay, that sounds nice. I'll be interested to hear what you've been up to."

I recall the time I had worked at that haberdashery during the summer when I was a teenager. The manager told me to "never talk past the sale."

"See-ya soon," I say.

I don't usually exercise late in the day but I feel I need a bit of a tune-up since my trip to Cairo had left my daily regimen in shambles. I do a few exercises and jump in the shower. Discipline in this area has always come easy for me since my early twenties when I decided to start exercising regularly. I hadn't run any significant distance since high school so I thought I would go down to the local junior college track and run a few laps. I started jogging, expecting to run a mile or so before resting. By the end of the second lap, I was exhausted. My brain told me to stop. "You're too tired," it said. "You're exhausted. You can't make it." That's when an extraordinary thought occurred to me: Don't argue with your brain but *do it anyway*. So, I did and have been doing so ever since. My internal dialogue went something like this:

Brain: Stop running, you're exhausted!
Me: Okay. (*Still running.*)
Brain: Really, I'm not kidding! You're too tired to keep running. You need to stop now before you keel over.
Me: You're right of course. I am much too tired to keep running. I am exhausted. I *really* should stop running . . . right now! (*Still running*)

And so it was I learned yet another life-lesson: Do what your brain tells you . . . unless you don't feel like it. But whatever you do, don't *argue* with your brain—it's entitled to its opinion.

I go to Linda's house and ring the doorbell. She opens the door and my eyes widen. She looks stunning in a bare shoulders low-cut black evening gown, matching handbag, with a small white pearl necklace draping her neck.

"Good evening," I say. "You look splendiferous."

"I bet you say that to all the women," says Linda, smiling.

"Only if they're named Linda and just happen to be molecular biologists."

I had made reservations at *Saint Michael's Alley* on Homer Avenue in Palo Alto. We decide to skip the entrée and order seared wild sea scallops and spring vegetables with grilled lemon butter and a glass of chardonnay to wet the palette. We share similar tastes in food and entertainment, generally. Could that be a good portend for our relationship? I'm looking forward to an enjoyable evening—just the two of us sharing the evening.

It's a weeknight so the restaurant isn't very crowded. We are seated next to a couple of empty tables. I'm glad we were able to find a private spot.

"Do you think we'll ever make any sense out of what happened in Cairo?" asks Linda.

"I have an idea but it's pretty bizarre."

The server comes by with our wine. "You go first," says Linda. "My idea is probably even more bizarre than yours."

"I don't think the Egyptian government had a clue what was going on," I say. "They gave Syed a cock-and-bull story and then retracted part of it later. I think they just made up a story about the Santa Fe Institute's involvement and when they realized it could be easily checked out, they told Syed it was part of an 'international effort.' I think President el-Sisi called the White House and after speaking to the president he was asked to keep everything under wraps."

Linda takes a sip of wine. "I think that's what happened too," she agrees. "I'd love to have a transcript of their conversation."

"That could tell us everything we're trying to figure out in a nutshell. The next time I speak to Syed I'll ask him if he can get closer to the inside story. He's connected. I know that. But here's the bizarre part: Syed told me of two tablets that are essentially the same going back three thousand and eight thousand years ago. "

"The same? What?" Linda isn't following.

I need to be clearer. "Both plates refer to King Zet who reigned about three thousand BC. One of the plates is just as we would expect. It refers to his reign which Syed told me took place about this time. The *other* plate also refers to the same king but the radiocarbon dating places its age closer to eight thousand BC, a five-thousand-year difference."

Linda is sitting back in her chair, relaxed but listening. "I almost thought of your parallel universe idea but both of the plates are here in our universe so that can't be. Can it?"

"You know what I think? I think something like that *is* happening, but not in a way we might imagine. I think information is being exchanged *between* universes. I know it sounds radical but the more I think about it, the more I realize that any conventional scientific approach isn't going to lead us to a correct answer. We are witnessing events that refuse a mainstream scientific approach."

"You seem to be saying that science doesn't work. That's heresy!" says Linda, mixing the serious with the silly. I know she is just trying to draw me out to better explain myself.

"We *knew* the Earth was flat, that the entire universe consisted of nothing more than the Milky Way, that it was static and so on . . . until we didn't. We've now figured out that these notions were false and replaced them with 'new and improved' ideas that give us the *correct* view. We're *pretty sure* we *probably* live in a 'multiverse,' consisting of all possible universes containing all of existence. Physicists didn't think this was such a good idea at first but now we do. It has great explanatory power. But we could be wrong. Probably not, but we *could* be."

Linda is now sitting up straight. "It sounds like you're saying we don't *quite* know what we're talking about. Is that what you're saying Jake—'all bets are off'?"

"No, what I'm saying is that what we know, what we can prove, works just fine until we are presented with new information that contradicts what we *knew*. We need a completely different way of looking at what is happening. I'm not suggesting we toss out the tools of science. They work well enough as long as the facts cooperate, as long as we're dealing with what we have always agreed was possible. But this is different. We're seeing things that are not possible according to the laws we know. We have to come to grips with the possibility that what we have discovered doesn't agree with our current models of how things should work."

"So, what's the bottom line? Are we living in a universe of higher dimensions? Can space even have more than three dimensions?"

"No, that's not what I mean. Any universe we live in requires three dimensions of space only. Adding another would muck up Newton's inverse-square law for gravitation and wreak havoc with our solar system. Subtracting would cancel out gravity as an attractive force. No, we're stuck with three. But that doesn't mean an explanation requires higher dimensions, even rolled up dimensions that we can't observe directly, which are allowed by string theory."

I'm not sure if Linda got all that. Smart as she is, this isn't her "neck of the woods." I wonder if I should complete my thought about what happened. Oh well, I might as well get it all out on the table.

"I think that in another world just like Earth there are people like us, maybe even identical. And somehow information is being swapped between our worlds. The people living there are ahead of us by about five thousand years." A crazy thought pops into my head. "We may have lived there—you, me, Syed and everybody else now living here on Earth and—"

"Wait a minute," says Linda. "You don't mean *we* 'we'?"

"Why didn't you say you needed to go to the bathroom? I can wait."

We laugh. I thought it was clever. "Not exactly. More like versions of us experiencing similar though not necessarily identical timelines."

"Are they living in another universe or our own?"

"Wow! That's interesting. I hadn't thought of that—two versions of us in the *same* universe. The odds of another parallel universe next to our own would be enormous but in our universe? I don't think I even know how to calculate such odds. But, hey, this whole thing is strange enough; we might as well throw that possibility into the mix."

"Are 'we' having dinner together in the other universe?" asks Linda.

"We had dinner a long time ago, about five thousand years or so, give or take."

"But that doesn't sound right," says Linda. "The way you explained it to me any 'doppelgangers' as you call them would be living at the same time as us. It's the decisions we make that allow for different timelines but they're all occurring at the same time. Isn't that what you said?"

"Yes, but I'm not talking about your 'run-of-the-mill' parallel universe. I'm suggesting a different evolution altogether, one that allows *staggered* timelines. Everything in our lives takes place normally as it does now but is shifted—in this case by some five thousand years or so. The 'Egyptians' on the other world reached King Zet's reign approximately five thousand years before they did on ours."

Linda's intellect has been kicking back, taking everything in. It now springs to life. "The three tablets contain advanced scientific information from the other world that coincides with our own time, give or take ten or fifteen years. But they reached this point five thousand years before us. How can others like us living on a planet identical to ours mimic the lives we're living now, doing so five thousand years *before* we were even born?"

Here goes nothing. "Jeremy told me the image of the cosmic background radiation suggests we may be living in a holographic universe. He didn't exactly put it that way; that was my inference from what he told me. But if true it could explain everything, at least to my satisfaction. The only problem is that the explanation won't sit well if it becomes public knowledge. I think that's why the governments are involved in a cover-up. Not all of them of course, but at least the United States and Egypt—probably others as well."

"Let's say we do live in a holographic universe," Linda says, trying to follow my train of thought. "What does that have to do with parallel universes and doppelgangers and staggered timelines? I don't see the connection. Aren't they different ideas?"

"Okay, I'll tell you, but you'll have to cut me some slack. Putting things together that don't fit isn't easy. It's not like forcing a square peg into a round hole."

"Just tell me already."

"The two worlds are not separate in the traditional sense, they're interconnected." I look at Linda scrunching her eyebrows trying to push my statement deeper into her brain. She doesn't say anything so I continue. "Just as information about objects falling into a black hole is stored as data in the surface fluctuations, all of our histories are embedded on the boundary of the multiverse. Your history and mine, as well as the histories of everyone and everything else, are recorded in the boundary—an encoded hologram. Everything—past, present, and future—is stored in this holographic boundary. The encoded information is complete; we are merely living a 'reflection' or 'projection' of a specific segment of the hologram."

This is where an analogy might help. "Imagine someone watching a movie—not you, someone else. You and I are in the movie along with everyone and *everything* else. You can imagine how huge the screen must be. The viewer is watching you and I having dinner. I'm telling you this idea about a holographic universe and it suddenly all clicks. We—you and I in the 'movie'—realize that our lives are *scripted*, that everything we have done, are doing, and will ever do has already been 'recorded.' We are of course startled by this discovery and, being the highly evolved beings we think ourselves to be, come up with a very clever argument as to why this can't possibly be the case. We argue that we *can't* be living out scripted lives because we are arguing about that very possibility and *that* can't be scripted. The person viewing us on the screen and listening to our argument then looks at a script she is holding which contains the *exact words we are speaking*."

Linda is deep in thought. Just as she is about to make a point the server comes with our meal.

"This looks good," says Linda, looking at her plate of scallops. "So, can we get a copy of this 'script'?" she asks. "It might be nice to know how this mystery finally turns out. Or is that not in the script?"

I can't tell if Linda has been taking me seriously or is just kidding. I take a couple of bites of scallops. "This is good," I say, giving the more difficult subject a rest.

Linda then gets back on track. "If information about everything that is ever going to happen is contained in a hologram at the boundary how is it that the plates referring to King Zet got 'switched' between worlds?"

"A glitch," I surmise. "It's a hologram; a 'program' of encoded information."

"Well, what could cause that?"

"I don't know."

Linda isn't satisfied. "How do objects get 'swapped' between universes? That doesn't make sense."

"It is Information—pixels if you will—that is being 'swapped.' " Another analogy comes to mind. "When people used to play vinyl records on a phonograph—and some still do—if the record was warped, the stylus would sometimes skip and you would hear the same lyrics again. The different timelines between our world and one very much like our own are separated by about five thousand years. The ivory tablets documenting King Zet's reign are proof of this. The hieroglyphs on the tablets differ somewhat but otherwise contain the same information. We are living the past from the standpoint of the other world and they are by now well into the future from our point of view. They were where we are now five thousand years ago. That's why the information contained in the three tablets Syed discovered is contemporary—for the most part—with our present timeline although reflecting ancient knowledge from their perspective."

"I feel like Alice trying to convince Tweedledee and Tweedledum that she's real and not just a character in the Red King's dream. But I'll try my best not to cry."

"It's not so bad," I say, "as realities go. It could be worse."

"I suppose it wouldn't be so bad if we could pick out our reality, like that guy in *The Matrix*," says Linda.

Now Linda is in my genre. "You mean Cypress?"

"I don't remember his name. But the point is, if I understand you correctly, we don't have 'skin in the game.' It's a fait accompli."

This isn't the time to convince Linda I believe there is a "silver lining" behind our apparent *free will* predicament. It would take too long and might lead to an argument, the last thing I want. I steer the conversation to safer grounds and we spend the rest of the evening relaxing and enjoying our meal together.

After I take Linda home, I walk her to the door and, before going in, she moves in closer and we kiss. We hold each other for a moment and then she turns around to go inside.

"Good night Jake, sleep tight."

"Don't let the bedbugs bite."

Walking to my car I think about how corny our comments were . . . perfect. I feel like Sally Field must have the night she won her Oscar for *Places in the Heart.* "You like me, right now, you like me!"

Chapter Twenty-One: Altered States

Syed calls me early in the morning and drops a bombshell. He has found "the smoking gun."

"Jake you must listen carefully. I have some very important news."

Syed is calling from his personal phone. Had he slipped up or dismissed his earlier concerns about surveillance? I am pleased he had taken the initiative, saving me from having to ask for his help—not that he would mind. I assume he has found out more about the governments' role in the attempted cover-up. I am not disappointed.

Syed's voice is slow, measured. "I met with a cabinet member from the government. He and I have known each other for some time. As we suspected, he confirmed that the government had been unaware of the documents before we acquired them."

Understandably, Syed hasn't given me the name of the government official. What else has Syed learned about the true nature of the documents and their origin?

"Did he have any idea where the documents came from or why they contained so much advanced scientific knowledge?"

"Yes, he said there was an encrypted message sent to President el-Sisi from the United States. He is aware of its contents but has not seen the document personally. The president spoke to him on Monday and a meeting was held with a few select members of the *Supreme Council of Information*. He told me the role of the council was very different from what the public had been informed."

This is the key that will finally clear everything up. My speculations will be confirmed or we will be presented with a different explanation. Either way, we will have an answer. "Did he explain what he knew about the communication from the U.S.?"

"He did tell me the documents were not from our time. He was however hesitant to say anymore. His caution is understandable. The information and the . . . implications, if widely known, would create a great deal of concern in the world community."

I know this is going to be a stretch but if Syed could find out this much, what not ask? "Syed, is it possible to obtain a copy of the message?"

"I do not know Jake. That would be very tricky. He has already compromised—"

"That's alright, Syed, I wouldn't want you to get into trouble with your government. Is there any other way we might be able to find out more about what happened? Do you have any ideas?" I'm grasping at straws.

"I will see what I can do."

I don't see any other options if Syed can't get closer to the truth. He is better connected than I could ever hope to be.

After I get off the phone with Syed, I check my messages and find one from Jeremy. He had sent me a copy of a bulletin issued to the members of the *International Astronomical Union*:

Technical Meeting Proceedings:
ASP Conference Series 100
Astronomical Review: Inter-world Communication
Holographic 2-D Information Grid.
IAU Technical Workshop
Paris, France
Eds. G. Farland, K. Simmons & F. Terrazas
ASP Conference Series
ISBA 2-42776-293-0

Jeremy appended a note: "Thought you might like to see this . . . Coincidence?"

The Union's mission is to "promote and safeguard the science of astronomy . . ." Countries the world over participate in professional research and education. The membership exceeds eleven thousand and their collaborations cover a wide range of issues. They would not, however, be the "go-to" organization to manage the scientific implications or impact of Syed's discovery. The bulletin made the meeting sound routine. There was no reference to any international or world crisis or suggestion that a recent discovery had prompted the conference.

I call Jeremy—"the gift that keeps on giving." I ask if it is possible to get more details on the conference, anything that might tell us what prompted the meeting or provide a more in-depth description of exactly what is under investigation. He says he will look into it. He must have been busy. He usually has more to say. It occurs to me that I am the benefactor of his generosity and haven't reciprocated. Maybe I could pick up a couple of tickets to a ballgame and invite Jeremy.

After thinking about Jeremy's text, it occurs to me: The international scientific community is being used surreptitiously to ferret out the possible meaning and implications of Syed's discovery. The governments aware of the find needn't bother concealing anything, except the truth. By requesting a colloquium to investigate the possible consequences of the find without revealing what happened—convening the conference to simply explore or investigate hypothetical *possibilities*—the governments could remain complicit, giving nothing away.

Would Linda come to the same conclusion? I send her the text and wait for her reply.

Amid the intrigue of government conspiracies and my attempts to find an explanation for what had happened, I have lost track of a key fact: there is an Extinction Level Event coming our way, an asteroid that could potentially wipe out all life on Earth. Jeremy's text hadn't mentioned any recognition by the *International Astronomical Union* of the ELE in their upcoming meetings. Were they unaware that the asteroid is expected to veer from its normal path, becoming a potential threat to our planet? If so, it would be up to the U.S. and Egypt to provide an appropriate response to the threat, informing other countries according to international conventions and, of course, politics.

Jeremy calls me late in the afternoon waking me from my power nap. I don't mind.

"Jake, guess what I found out?"

"What have you found?" I ask, hoping Jeremy will cut to the chase.

"I couldn't begin to explain the synopsis over the phone," says Jeremy, "but I wanted to give you a heads-up before I sent you a copy. I can't think of any reason why they would be looking into such weird phenomena unless they are getting direction from somewhere. I don't think they would be exploring such wild scenarios on their own. It's not in their wheelhouse."

Jeremy sends me a text of the synopsis and, after I thank him for his help, I ask if he would like to take in a game some time. Jeremy is up for it and we end the call. I immediately look up the word "wheelhouse": a baseball idiom meaning "within the zone that is most advantageous for a batter to hit a home run." It figures—Jeremy, the quintessential astrophysicist, and baseball aficionado.

The information in the synopsis provides strong evidence in support of my hypothesis. It goes further, however, by introducing the possibility that the entire multiverse, or meta-universe—*everything that exists*—is "hidden" behind the event horizon of a single black hole. All "parallel" or "alternate universes," including our own, are "encoded" in the black hole's two-dimensional event horizon: We live in a hologram.

The synopsis outlines a possible cosmology, speculative, and unproven as far as the IAU is concerned. They have no way of knowing that their collaborative efforts are being engineered, manipulated by real concerns unknown to them. To the IAU, it is simply an exercise exploring possibilities, probing the limits and implications of new and radical ideas.

I know better. Our lives, our 3-D universe—*everything*—is a projection of a flat, two-dimensional surface far removed in time and space. In this singular, all-encompassing, hologram all reality is connected. I now have an explanation of what happened and how events will, necessarily, resolve. The "glitch" or perturbation in the two-dimensional hologram occurring on the surface of the event horizon, a frequency anomaly, caused the transposition of *information* between parallel universes. The material wasn't being swapped, "pixels" were—"diffractive pixels" . . . holograms containing "voxels" ten trillion, trillion times smaller than a single atom; what physicists refer to as the "Planck scale."

Timelines were disrupted, information "exchanged," and reality altered. When the hologram adjusts, like tectonic plates after an earthquake, timelines will be restored and reality will, once again, revert to "normal." Memories? If we are projections, three-dimensional images encoded in a two-dimensional surface, our "memories" are also projections and they, too, will synchronize with our former reality before the disturbance. During the interim, we interact with the world as it is—disrupted by the mysterious appearance of ancient Egyptian tablets containing modern scientific knowledge—unaware that our former reality will be restored, dialed back, to a time before the "quake" occurred. When the realignment will happen is an open question.

Linda sends me a text asking me to call her. She has my text about the meeting of the IAU that Jeremy had sent me but not a copy of the synopsis. Should I send it to her? The conference shouldn't raise any red flags. It might not be a conventional or mainstream subject for the institute to cover but, other than its novelty, it isn't completely unorthodox either. I doubt the event will get any substantive press coverage. I can't think of any reason why it might come across Linda's radar. Of course, Linda being Linda, she will probably pursue the matter one way or another.

I call Linda.

"Hey."

"Hi Jake," she says, dropping our usual greeting. "This bulletin . . . the subject of the meeting can't be a coincidence. There must be a connection. What do you think?"

I decide to play it low key. "It may not be a coincidence but I don't think they're *in on it*. I think they were given the agenda and just following directions. If they knew what we do the subject wouldn't be on their schedule as a routine meeting, it would be international news."

"Or they would keep their investigations and report secret altogether," says Linda.

"Syed says he will do what he can to find out more but I don't think he's going to find out anything he doesn't already know. It's up to the U.S. and Egypt and whoever else they are willing to share the information with to determine what to with the documents." I can't bring myself to tell her that, sooner or later, it won't make any difference. We will be back to square one where we started—no tablets, no mystery . . . no Linda and me . . .

Linda snaps me out of my musings. "Is there any way to find out more about the meeting? Maybe Jeremy could get some inside info."

"Well, what if we did know . . . everything? There's not much we can do about anything now anyway."

"So, you're saying we should just give up?"

I can't see where this line of thinking is taking us. What are we supposed to accomplish? We don't have the information, and if we did, what could we do? Publish the documents? Then everyone else would know. Would that be good or bad? I think of the Sufi story, "We will see."

"I don't see us getting our hands on the documents again," I say. "Even if we did and we published them, would that be the right thing to do? I don't know."

"So, you *are* saying we should drop it," says Linda. Then, pausing for a moment, she concludes: "Maybe we should."

Linda seems resigned. I don't get the impression she feels rejected or defeated, just resigned to the situation. I don't mention the synopsis. I tell Linda I will let her know if Syed comes across anything that would *change* our assessment or prompt any further action on our part.

I think about how my new understanding of the meta-universe will affect my life and realize it won't have any effect at all; none whatsoever. Once the holographic anomaly self-adjusts, I and everyone else will no longer have any memory of what happened. The tablets and documents will never have existed. My version of reality will have been amended along with my memory of the discovery of the Egyptian plates and my trip to Cairo. The agents who had transported the tablets would not have died in the car crash and subsequent explosion.

As for my relationship with Linda, I will have no recollection of the time we spent together. Giving up those memories will not affect me as I will be unaware that they had been erased. I will end up where I was before we became more involved. Does that mean our relationship would grow as it has already or fail to progress at all? The circumstances will be different—no Cairo, no *adventure*. Would that matter? "We will see."

Is there any way for me to know, for sure, if I have lived through these events before? Am I trapped in a *temporal causality loop*, forever repeating the same timeline over and over again? How could I know? My assessment of the events Syed, Linda, and I witnessed is speculative and unprovable, and, ironically, will remain so, *if* true—as the *proof* must disappear if my hypothesis is correct.

How do we understand the universe and our place in it? The media, politicians, and the government lie to us, sometimes blatantly, with impunity. Is what we have been told about how our universe works by the scientific community the truth? It seems a strange question for me to ponder as I am a member of the very same community of scientists ostensibly dedicated to the search for truth.

When I first learned that our universe was "fine-tuned" for life and our Earth is situated, coincidentally, in the "Goldilocks Zone"—not too close or far from the sun—I thought it was remarkable proof that the universe was made for us. Then I learned about the *Copernican Principle* or *The Principle of Mediocrity* which tells us that we are not privileged observers: the sun and Earth are not in a central position. The conditions required for life—even intelligent life—are readily available the theory explains. There are millions of Earth-like planets and ours is but one of many scattered throughout the cosmos. Further, since there is an infinite number of universes in the meta-universe, finding one in which the physical constants of the universe are "fine-tuned" for life—our universe—is unremarkable.

Was the idea of "mediocrity" invented to throw God out of the equation? Is the theory of "eternal inflation," responsible for the multitude of universes we need to demote our own to *mediocre* status, a valid theory? Based upon our discovery and the anomalies we had uncovered; it would seem so. The idea of a black hole containing a single universe or the entire meta-universe is, in theory, conceivable. Did our discovery resolve these questions or present new ones that we may be forever unable to answer?

Ironically, our "little neck of the woods," the *Milky Way*, will one day, once again, be the center of the universe, or, more technically, our metagalaxy comprised of the "Local Group" all merged into one, will be all that anyone living in the former *Milky Way* could observe. Just beyond the unfathomable reach of time, all other galaxies will have receded beyond our observable horizon, and light from them will no longer reach us. *If* intelligent creatures were to gaze out from this singular metagalaxy—having lost all records documenting a much larger universe—all they would see, using their most powerful telescopes, is our quaint little region of space. A place we like to call home.

We are *presented* with a scientific view, a model of reality, and are asked to accept it. Do we have a choice, any say in what constitutes the "correct" view? Are we permitted to present alternate views of reality? Yes, of course, as long as our ideas *conform to scientific orthodoxy and previously agreed-upon protocols for establishing scientifically verifiable proof of our arguments*; otherwise, no—we're "crackpots" whose ideas should be relegated to the dustbin, along with all of the other crazy notions, such as alchemy and witchcraft. But who are the gatekeepers of "scientific orthodoxy"? Who decides which ideas are meritorious? The scientific community, of course! Who else? We come up with ideas according to procedures we formulate, applying "peer-reviewed" critiques—by a cabal of like-minded and formerly indoctrinated scientists—and give a "thumbs-up" or "thumbs-down." Non-members of the inner sanctum need not apply; it's a closed-loop from start to finish.

It has always made perfect sense to me that scientists should be in charge of "scientific" ideas. Bureaucrats are in charge of bureaucracy, educators in charge of education, and so on. But what if *unorthodox* ideas have merit but are rejected out of hand by the scientific community? It is the price we pay to retain a measure of control over the scientific process. Otherwise, we would be wasting valuable time chasing one "hair-brained" idea after another. And yet . . . and yet, I can't shake the feeling that a new paradigm is throwing logic out the window, that my view of reality, as a scientist, is being called into question.

If our universe grew exponentially or shrank to microscopic size instantly—atoms, quarks, everything—how could I detect that it had even occurred? If everything were instantly frozen, and, just as quickly "restarted" would it make any sense to speak of "how long" it had remained motionless? Wouldn't time have been "frozen" as well? How well do we understand anything?

Contemplating the bizarre aspects of the reality, I begin to question whether or not I am real—a three-dimensional, corporeal, entity, and not an artificial construction or projection, an imposed reality manufactured by a two-dimensional hologram on the surface of an event horizon of a black hole containing *everything*. Were Tweedledee and Tweedledum right after all? Or am I losing my mind?

René Descartes thought he had it nailed with his, "Cogito ergo sum"—"I am thinking, therefore I exist." I'm not buying it. All it means to me is: thinking is occurring. No agent is required; at least not a physical, human, agent. We tell ourselves we are alive and exist by virtue of our senses; they give us feelings, sensations. If they are real, we say, then we must be as well because we are the ones experiencing the feelings. But is it really any different than Descartes' "thinking"? If I had been born without my senses, by what means would I confirm my existence? Since I would be incapable of thinking, at least as we believe we understand the process, I would have unthinking, internal, mental states—but how could they confirm anything? I feel another soft migraine coming. Time for another power nap. Maybe I will drift off into dreamland—that's where reality . . .

Linda wakes me from my nap.

"Hey," I say. "Looking forward to the weekend?" I ask, hoping to encourage an invite.

"Hey Jake," says Linda. "I could use one. I'm still trying to get back in the groove. I wanted to ask you if you would like to come over for dinner Friday?"

"I'll have to check my social calendar," I say, "but I think I can squeeze it in."

Linda laughs. "How do you feel about calamari?" says Linda, knowing full well I can't stand it.

"With a side of chocolate ants? My favorite! How'd you guess?"

"Okay," says Linda, getting a little more serious, "how about Italian, around seven?"

"Now you're talking," I say. Wanting to put the lid on the invite, I quickly move on to another topic. "Have you been able to catch up yet?"

"No, I'm still *getting* caught up. I'll need to complete some papers over the weekend but I'll make room."

"As long as you make room for me."

"I always have room for you, Jake."

Linda's invite is just what I need. After her call, I begin to think about what really matters: How Linda and I feel about each other and what I need to do to get my life back on track.

Chapter Twenty-Two: Destiny

Ask a neurologist what love is and you will be told it is all chemistry; a philosopher, obligation; a romanticist, obsession. Is it any of these, all of these, a mix, or none? It is a feeling, and, as such, it has a neurochemical basis. It invokes passion—even obsession—and, therefore, can be properly understood from a romanticist's point of view. And, finally, if it endures, it will entail a sense of obligation. But is love a proper subject for analysis?

It was once believed that our investigation of nature, of the cosmos, would extract the wonder and excitement, the grandeur out of the natural world. If you ask most scientists, however, they will tell you just the opposite—knowing how nature unfolds her secrets adds to the sense of wonder and excitement. Knowing that the size of the moon and its distance from the Earth allows for a perfect solar eclipse, and that, eventually, the moon will be too far away to accommodate this most perfect of celestial phenomena, makes me appreciate the time and place in which I live more, not less.

Understanding the neurochemical substrate of love—dopamine, phenylethylamine, and norepinephrine—tells us what is happening from a physiological standpoint, but does nothing to diminish our feelings, our sense of longing and attachment. It may be a Darwinian trick to propagate our species, but so what? We eat because we're hungry and food tastes good. The fact that our eating has *survival* value doesn't make the food more or less tasty.

That my "chemosensation" responds to Linda's pheromones and the genes of my Major Histocompatibility Complex signal "diversity of antigen presentation" is good to know. Her sense of humor, intellect, curvaceous body, dark blue eyes, shimmering brunette hair, and beauty are not to be overlooked. But, as Bogart said in *Casa Blanca*, taken all together, they "don't amount to a hill of beans." I love her because I love her; I don't *want* to analyze how or why I do.

I pick up a bottle of *Carignane wine* for dinner and hand it to Linda when I arrive.

"Hi Jake," says Linda, taking the bottle and giving me a quick kiss. "This is perfect. I'm making Lasagna. Come in."

"You're sure the wine is okay?" I say, fishing for a compliment on my choice.

"Yes, Jake, it's just fine; except for the 'e' on the end of the name. It's too California 'ish.' "

"What do you mean?"

"Nothing," says Linda. "I'm just teasing you. It's perfect like I said. Sit down. I'll be in in a minute."

I sit down in the living room on a sofa in front of a window looking out onto the back yard. Except for the kitchen, the lighting is subdued. Linda's home, like mine, doesn't have a lot of clutter. There are a few paintings, the furniture is transitional, crème-colored sofas with contrasting throw pillows, a square coffee table with an open shelf.

"This coffee table is pretty cool," I say, throwing anything out to start a conversation.

"They call it 'The Crate.' " says Linda. "It comes in handy when I need to do some work at home and need to relax at the same time. I can keep some of my papers handy where I can get to them easy."

Linda comes over and sits down next to me on the sofa. "I was going to put some music on. What do you like? I don't know if I have it but—"

"Anything you like is fine," I say. We sometimes listen to "soft" music when driving around together, going to lunch or dinner.

"I've got some hard rap if you like," says Linda, immediately laughing. I'm relieved. "How about Nora Jones . . . never mind—"

"No, that's alright. I've listened to her music. It's pretty good. What was that one? 'I Don't Know Why'?"

"You like that?"

"Sure."

"Okay," says Linda. "I'll select it."

I've never liked loud music, even as a teenager. It just sounds like noise to me. Growing up, my friends used to tease me about it. They knew the lyrics of most songs forwards and backward. It was never my "thing."

"You never told me what kind of music you like," says Linda.

"I never got into music growing up. I can listen to pretty much anything as long as it isn't loud . . . or rap. We both like Nora. What else do you like?"

"I don't have a preference. I'm with you; I don't like loud music either. I listen if I'm somewhere music is playing but I don't buy and hoard it like a lot of people."

"We're alike in a lot of ways," I say.

"You never talk about TV shows you watch," says Linda.

"I seldom watch TV unless there's a specific show I like and then I'll usually 'binge watch.' "

"Me too. What do you binge on?"

"I usually wait until I see something good I think I'll want to watch and then I consume the whole series. I tape some old shows and watch them sometimes. Like the older series *House* and a few other programs. I liked the *Downton Abbey* series. The movie was pretty good too."

"I like the way it captured the history back in the early nineteen hundreds and moves through the war and so on. And the Crawley family isn't depicted as aristocratic snobs but genuinely nice people."

"I don't think some people are happy with that," I say. "Rich people are supposed to be greedy and evil. Some are, but not everyone who has money is a jerk. Anyway, it's fiction and covers a very different time. I just find the story and the characters interesting. Too bad some of them had to flake out and go their own way. Got a chance to make some money somewhere else I guess."

I'm comfortable with Linda. Not "like an old shoe;" it's different—we are *meant* for each other. It is at that moment I realize it doesn't matter if I have to start over from scratch to win Linda's heart. We will find each other again, fall in love. It is . . . *inevitable*.

"I think diner's ready."

We go into the dining area just off the kitchen. I ask Linda if I can help but she tells me to have a seat. I sit at the end of the dining table and Linda brings our meal over and sits next to me.

"I forgot the wine," says Linda. "I'll be right back."

I don't want wine. Well, I do; that's why I brought it—but not tonight. Tonight, I want both of us to be sober. It's a strange thought for me. Nothing wrong with a glass of wine with a good meal, especially when I'm with the woman I love. But something compels me to refrain. "I know I brought the wine but—"

"Is everything alright?" asks Linda. "You're not feeling well?"

"No, I'm fine. I just—"

"That's okay," says Linda. She probably thinks it strange but acts as if it doesn't matter. "I can get some tea if you like."

"That would be good."

"I'll join you. I don't feel much like wine tonight anyway."

Linda has a knack for doing and saying the right thing at the right time. After we finish our meal, we go into the living room and enjoy a late-night coffee. It will keep me up but so what. I don't have anywhere else to be and don't want to be anywhere else.

"Do you like children?" Linda asks. Just like that—out of the blue.

"Sure," I say, "I love kids, but I can't say I've given it much thought lately." I am off guard. I don't know what to say. "What made you bring that up? I mean, I don't mind, if you want to talk about it."

"No, I just think about it sometimes. I don't know what made me say it. I just feel like I can talk to you—"

"Linda, you can talk to me about anything anytime. I *care* what you care about. I guess I'd be in okay shape when the kid reaches his teens. I could probably still play a good game of catch."

"Congratulations!" says Linda. "It's a boy!" We laugh at my presumptuousness.

"I could handle having a 'Daddy's girl,' " I say. And then, being the socially inept Cretan, I say, "Maybe we could have three children, one of each."

Linda laughs. "Where did you get that one?"

"It's something I heard Jerry Lewis say once."

"I'll settle for twins," offers Linda, "a boy and a girl."

"Well, I'm glad we got that settled," I say. "But shouldn't we get married first? Or does that matter anymore?" Linda thinks I'm kidding.

"There's only one thing we haven't discussed," says Linda.

"What's that?"

"Where we're going on our honeymoon."

Am I letting this go too far? I am "all in," but is Linda? I have to find out.

"Linda, I don't want to play games. You must know how I feel—"

"I love you, Jake."

That was it, the four words I had been waiting for. "Linda, I can't remember when I haven't loved you. I just didn't know . . ."

Linda moves closer. We kiss, passionately, holding each other, our desire inflamed. We get up from the sofa and walk into the hall leading to Linda's bedroom, stopping a moment for another quick embrace, a kiss. Just as we reach the doorway to Linda's room, I stop. I remember our time in Cairo and the promise I had made if Linda ever again invited me into her room and was stone-cold-sober.

Linda takes a step into the room, turns around facing me, and says softly, "Aren't you coming in, Jake?"

"Wild horses . . .

CPSIA information can be obtained
at www.ICGtesting.com
Printed in the USA
BVHW041220181021
619204BV00012B/166